Lecture Notes of the Institute for Computer Sciences, Social Informatics and Telecommunications Engineering 495

The LNICST series publishes ICST's conferences, symposia and workshops.
LNICST reports state-of-the-art results in areas related to the scope of the Institute.
The type of material published includes

- Proceedings (published in time for the respective event)
- Other edited monographs (such as project reports or invited volumes)

LNICST topics span the following areas:

- General Computer Science
- E-Economy
- E-Medicine
- Knowledge Management
- Multimedia
- Operations, Management and Policy
- Social Informatics
- Systems

Javid Taheri · Massimo Villari · Antonino Galletta
Editors

Mobile Computing, Applications, and Services

13th EAI International Conference, MobiCASE 2022
Messina, Italy, November 17–18, 2022
Proceedings

Editors
Javid Taheri
Karlstad University
Karlstad, Sweden

Massimo Villari
University of Messina
Messina, Italy

Antonino Galletta
University of Messina
Messina, Italy

ISSN 1867-8211 ISSN 1867-822X (electronic)
Lecture Notes of the Institute for Computer Sciences, Social Informatics
and Telecommunications Engineering
ISBN 978-3-031-31890-0 ISBN 978-3-031-31891-7 (eBook)
https://doi.org/10.1007/978-3-031-31891-7

This Springer imprint is published by the registered company Springer Nature Switzerland AG
The registered company address is: Gewerbestrasse 11, 6330 Cham, Switzerland

Preface

We are delighted to introduce the proceedings of the fourteenth edition of the European Alliance for Innovation (EAI) International Conference on Mobile Computing, Applications and Services (MobiCASE 2022). This held online November 17–18, 2022, brought together academia and industry to discuss opportunities and issues of mobile applications and services. This year, apart from mobile computing application and services, MobiCASE 2022 focused for the first time on energy efficiency.

The technical program of MobiCASE 2022 consisted of 7 full papers, including 2 invited papers, in oral presentation sessions at the main conference tracks. The conference tracks were: the Main track and the Invited Papers Track. Aside from the high-quality technical paper presentations, the technical program also featured one keynote speech given by Giancarlo Fortino from University of Calabria, titled "Pushing Intelligence to the Edge of Internet of Things: A new Paradigm enabling Next-Generation Smart Systems of Systems".

Coordination with the steering chair, Prof. Imrich Chlamtac from University of Trento, was essential for the success of the conference. We sincerely appreciate his constant support and guidance. It was also a great pleasure to work with such an excellent organizing committee team for their hard work in organizing and supporting the conference. In particular, the Technical Program Committee, led by our TPC Co-Chairs, Bahman Javadi, Mohammad Reza Hoseiny Farahabady and Armando Ruggeri, have completed the peer-review process of technical papers and made a high-quality technical program.

We strongly believe that the MobiCASE conference provides a good forum for all researchers, developers and practitioners to discuss all science and technology aspects that are relevant to mobile computing. We also expect that the future MobiCASE conferences will be as successful and stimulating, as indicated by the contributions presented in this volume.

Javid Taheri
Massimo Villari
Antonino Galletta

Organization

Steering Committee

Imrich Chlamtac   University of Trento, Italy

Organizing Committee

General Chair

Javid Taheri   Karlstad University, Sweden

General Co-Chairs

Massimo Villari	University of Messina, Italy
Antonino Galletta	University of Messina, Italy

TPC Chair and Co-Chairs

Bahman Javadi	Western Sydney University, Australia
MohammadReza HoseinyFarahabady	University of Sydney, Australia
Armando Ruggeri	University of Messina, Italy

Sponsorship and Exhibit Chair

Christian Sicari   University of Messina, Italy

Local Chairs

Antonino Galletta	University of Messina, Italy
Massimo Villari	University of Messina, Italy

Workshops Chair and Co-Chairs

Edelberto Franco	Federal University of Juiz de Fora, Brazil
Márjory Da Costa Abreu	Sheffield Hallam University, UK
Lorenzo Carnevale	University of Messina, Italy

Publicity and Social Media Chair

Eirini Eleni Tsiropoulou   University of New Mexico, USA

Publications Chair

Giuseppe Di Modica   University of Bologna, Italy

Web Chair

Mario Colosi   University of Messina, Italy

Technical Program Committee

Armando Ruggeri	University of Messina, Italy
Javid Taheri	Karlstad University, Sweden
Antonino Galletta	University of Messina, Italy
Massimo Villari	University of Messina, Italy
Eirini Eleni Tsiropoulou	University of New Mexico, USA
Edelberto Franco	Federal University of Juiz de Fora, Brazil
Mario Colosi	University of Messina, Italy
Christian Sicari	University of Messina, Italy
Giuseppe Di Modica	University of Bologna, Italy
Auday Al-Dulaimy	Mälardalen University, Sweden
Muhammad Usman	Karlstad University, Sweden
MohammadReza HoseinyFarahabady	University of Sydney, Australia
Ahmad Taghinezhad	University of Tabriz, Iran
Ayeh Mahjoubi	Karlstad University, Sweden
Lucas Frank	Federal University of Juiz de Fora, Brazil
Ahmet Soylu	Oslo Metropolitan University, Norway
Skhiri Sabri	EURANOVA, Belgium
Thomas M. Prinz	Friedrich Schiller University Jena, Germany
Dragi Kimovski	University of Klagenfurt, Austria
Ilir Murturi	Technische Universität Wien, Austria
Javier Berrocal	University of Extremadura, Spain
Dimitrios Kallergis	University of West Attica, Greece
Grzegorz Sierpiński	Silesian University of Technology, Poland
Juan Manuel Murillo Rodríguez	University of Extremadura, Spain
Uwe Breitenbuecher	Stuttgart University, Germany

Contents

x   Contents

Mobile Computing

Configuring Unconnected Embedded Devices with Smartphones

Peter Barth[✉], Nicholas Linse, and Rüdiger Willenberg

Hochschule Mannheim, Paul-Wittsack-Straße 10, 68163 Mannheim, Germany
p.barth@hs-mannheim.de

Abstract. We configure embedded devices with a smartphone via NFC using an open, platform independent protocol presented in this paper. A textual device specification defines the types of configuration values for a specific device and integrates the device into the configuration system. The specification needs to be provided by the embedded developer. It is translated into a C library that enables configuration value access, as well as *blob* that contains the compressed configuration metadata. A generic smartphone application interprets the metadata and configuration data read via NFC and allows the modification of the values according to the device specification encoded in the metadata. The modified configuration data can be stored, shared or transferred back to the embedded device. None of the configuration steps need an internet connection, which means data is kept private. Combined with the open protocol and the generic app, this ensures that embedded devices will not become obsolete through vendor decisions, as happens frequently with devices dependent on configuration via cloud services. Embedded developers only need to implement raw read and write binary access to an NFC storage device. The generated artifacts allow to transform that data into an easy-to-use data structure. A prototype system using a fully functional tool chain, a generic Android app and a single-board computer simulating an embedded device has been implemented and evaluated.

Keywords: embedded device configuration · NFC · smartphone app

1 Introduction

Increasingly, embedded devices are connected to the cloud to configure them via the web or a proprietary smartphone application as shown in Fig. 1a). For the device manufacturer, this has three advantages. First, the hardware cost and complexity of the embedded device can be reduced, as no expensive input and output hardware such as displays and keys need to be included. Second, the manufacturer benefits from the configuration data that is collected remotely. Third, customers get locked into the ecosystem of the manufacturer, making it harder to switch to other brands. As devices often stay connected, even if that is not necessary, the vendor may collect usage data to gain insights into the customer

© ICST Institute for Computer Sciences, Social Informatics and Telecommunications Engineering 2023
Published by Springer Nature Switzerland AG 2023. All Rights Reserved
J. Taheri et al. (Eds.): MobiCASE 2022, LNICST 495, pp. 3–17, 2023.
https://doi.org/10.1007/978-3-031-31891-7_1

4   P. Barth et al.

base. However, privacy concerns make both consumers and business customers hesitant to freely provide usage and even configuration data to their suppliers. Why should a lamp know the WiFi-Password (credentials)? Why should a coffee machine vendor know how much (too much) coffee we drink (usage data) and store that in the cloud? Why should they even know that we drink our coffee stronger in the morning (configuration data)? Why should washing machines, power outlets or heating systems need to be connected to the Internet? At least in Europe, the principle of data avoidance and minimization [15, Chapter II, Article 5] is enforced through the General Data Protection Regulation (GDPR) and most embedded devices could run fine without communicating with a cloud server. In addition, personalized data is sensitive according to the GDPR [15, Chapter I] and collecting it may make embedded device vendors subject to severe penalties for violations when processing, distributing or simply storing that data with the hyperscalers, especially if servers are located outside the European Union [19, Chapter 4.1]. All embedded device vendors are keen on reducing hardware costs and many might even give up data collection to avoid potential GDPR-violations altogether, provided they retain the cost savings and may even receive an influx of more data conscious customers. Furthermore, strategy changes or vendors going out of business may make required remote services defunct and thus, otherwise well-working devices become obsolete. Without vendor lock-in, embedded device manufacturers can offer greater sustainability and are more attractive to to buyers with privacy and environmental concerns.

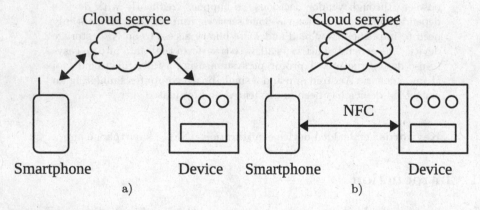

Fig. 1. Configuration of embedded devices via cloud service or direct communication

We propose to replace the omnipresent mid-to-long range wireless chip in embedded devices with an NFC chip [12] and to configure the device with an NFC-capable smartphone and a generic app as shown in Fig. 1b). Thus, the device vendor profits from the reduction in hardware and development cost as well as the easy-to-use configuration process. While there is no need to store data in a cloud, this may still be offered and even easily integrated. But in order to use a device, no registration or other lock-in is necessary. Therefore, devices can be used and configured beyond their official support lifetime or the lifetime of remote services. For many embedded device vendors, it might even be attractive that no data needs to be stored or processed, and that the customer provides the necessary input/output hardware, the smartphone, themselves. With a generic app, the embedded device vendor has no need to provide a proprietary app, while that still remains an option. Another advantage of remote configuration is that there is almost no configuration logic on the embedded device itself, which simplifies the firmware. To foster adoption of a disconnected approach without remote services, embedded device developers are provided with a tool chain that translates a human-readable representation of the configuration data specification into artifacts that are easily integrated into a firmware without further processing. For embedded devices that is a C struct holding the configuration data. What remains for the embedded developer to implement is the reading and writing of binary data to the NFC chip. The generic app can display and manipulate any data based on the metadata that is generated by the tool chain. The interaction and visualization in the app is dynamically created from the metadata and can be personalized. Data sets can be stored locally for later retrieval, reuse, modification and, optionally, sharing. In Sect. 2, we discuss related work for generic configuration data handling. The possible configuration options and its interaction are collected in Sect. 3. In Sect. 4, we provide an overview of the proposed solution architecture, consisting of the specification of the configuration data in Sect. 5, the tool chain in Sect. 6 and the generic app in Sect. 7. With a test bed using a Raspberry Pi we validate and evaluate the approach in Sect. 8. We conclude in Sect. 9.

2 Related Work

Most apps that rely on NFC are limited to process static information from a tag, such as a URL [3] to be redirected to a Website. A more interesting use is sharing initial pairing information over NFC to establish connections between devices over other communication methods [11]. There are projects which use NFC for general peer-to-peer communication [10], but no widespread applications. Using NFC for device configuration [1,7,16] is mostly proprietary and limited to configuring one specific device. Standardization efforts to offer a common approach to device configuration via NFC are not known to the authors.

Generic configuration of, and interaction with, devices is more widespread in the personal computer space. A good example is the Universal Serial Bus (USB). For classes of devices [18], for example input or audio devices, there are generic drivers that offer limited interaction and configuration of all devices of that class. Specific devices have specific features and consequently require specific configuration, which is available in a device descriptor offered to the host [17, Chapter 5]. For classes of devices, the specific device may choose from a selection of data types and additional attributes and thus follow a similar idea as presented in this paper. Likewise, such an approach can be used to configure Peripheral Component Interconnect (PCI) devices [2,14]. This generic approach to the configuration of, and interaction with, devices leads to a reduction or elimination of device-specific driver code. A similar approach has also been used to handle communication of Internet of Things (IoT) devices with their backend server generically [9] and to control peripherals of programmable logic controllers over a generic interface to increase program portability [5].

A declarative approach to building user interfaces is typical for app development. Many UI frameworks allow for the declaration of layouts (XML on Android, XAML on .Net) and there are efforts to provide generic collections of UI elements [8] to compose complex interfaces with. Most of these declarations are integrated into the app at compile time. Generating interfaces at run-time through received information is rare. In [4] generating interfaces from a declaration at runtime is discussed to enable device specific adaptations of the interface. Another example would be Facebook's Lite app [6]. This application is intended for devices with low processing power and thus all expensive processing is done in the cloud, which only sends a description of how the final interface should look. The application constructs its interface directly from this description.

3   Configuration of Embedded Devices

A wide range of devices, including coffee machines, time-controlled power outlets, central heating units and a signal generator for use in a laboratory, shall be configurable with the approach presented. All configurations can be reduced to a combination of few data types with additional constraints as well as typical, related interactive widgets. The most-used data types are integral and real numbers, selecting one or more out of many possible options and structured combinations such as date and time. As shown in Table 1, for example, a coffee machine needs a selection for the kind of coffee, the strength, the amount of water and a few additional features. To configure these properties, at least input of integral numbers and selecting 1:n or m:n predefined options is necessary.

Typical interaction patterns for 1:n selection might be some radio buttons for a few options or a dropdown list for more. Some people might prefer to specify the strength with a slider, others might want to input an exact number of grams of powder. For the time-controlled multi power outlet, a setting for the operating time range is needed. In addition, we need to select which of the sockets should be operated, an m:n selection, which could be conveniently selected via check marks. Additionally, not just one but multiple active time ranges and socket selections should be possible. As multiple similar time ranges are possible a way to store multiple data sets of a collection of settings is needed.

The signal generator has settings for the type of signal, frequency, and amplitude. Once again, these settings should exist separately for each output. For the signal type, a 1:n selection is needed, while the other settings can be directly input as numbers, as these might need to be exact. Because frequency or amplitude can only be set in a range that is supported by the device, this must be reflected in the settings, so numbers must support numeric limits for the input values. A generator might only support a range of discrete values, which are not

Table 1. Device examples and used data types

device	feature	data type	interaction
coffee machine	type	Integral number	selection (1:n)
	strength	integral number	number input/slider
	amount	integral number	selection 1:n
	additions	array of m booleans	selection m:n
	time control	date+time+boolean	special
	time control select	boolean	switch
power outlet	on/off	boolean	switch
	multiple profiles	array of complex settings	(repetition)
	profile active	boolean	switch
	time range	date+time	special
	active sockets	array of m booleans	selection m:n
signal generator	on/off	boolean	switch
	output	integral number	selection 1:n
	signal type	integral number	selection 1:n
	frequency	integral number	number input/slider
	amplitude	real number	number input/slider

on a linear scale. In this case, a list of the supported values needs to be defined, which is then used to constrain the input number. Thus, storage of the data is straightforward, while the preferred interaction style may vary depending on device and user preference.

The configuration will be received, stored and read mostly on low-end embedded hardware. Thus, hardware requirements and code complexity should remain low, the language of choice is C. The embedded developer should focus on reading the configuration data from a struct to take action based on the contained values. The used protocol and data structures need to be small, simple and natural for an embedded developer. Thus, no dynamic memory allocation shall be used, as this is often not available on low-end microcontrollers and would limit applicability. The communication protocol should provide very low overhead and thus be binary. Converting between communication data and the storage in a struct should be provided by a generated library. Due to the diverse embedded environment and hardware specialties, not to say bugs, accessing and storing binary data to/from an NFC chip cannot be abstracted away and need to be provided by the embedded developer for this specific NFC tag and board.

For the mobile app, we start with widely available Android smartphones. They have the required performance and interfaces that users are accustomed to. NFC is used for communication as it is increasingly openly available on Android smartphones and cheap as well as easy to deploy on embedded devices. Communication can be established by easy and intuitive touching without the need for complicated pairing processes in related technologies such as classical Bluetooth. The need for physical proximity replaces special security protocols.

However, NFC connections are not suitable for transmitting large amounts of data or to hold a connection for a prolonged time. As we only concentrate on the configuration of devices, not direct interactive control, and the transmitted binary configuration data is small, this does not hinder effective usage.

4 Architecture

Configuring an embedded device over NFC, as depicted in Fig. 2, is triggered by bringing the smartphone into proximity of the device. In one data transfer, the smartphone receives and extracts both data and metadata from the embedded device. Based on the metadata, a user interface is dynamically generated, which allows to change the values according to the specified restrictions found in the metadata. After changing the values, the user may decide to write the changed values back and touches the embedded device again. This again triggers transmission of both data and metadata, but this time back to the device. On the device the transmission is registered, and the newly written values are provided to the firmware. Currently, there is no cryptography involved for authentication, integrity, or secrecy of configurations. Proximity is sufficient to change values. It is advisable, that the firmware developer has to check that the metadata is unchanged, as there is no portable way to only write part of the structure back to the NFC chip. Thus, if the metadata is modified, the embedded developer needs to initiate restoring the metadata from a backup ROM, if it has changed. In addition, the firmware developer has to check that the data itself is sensible, which is helped by generated C code.

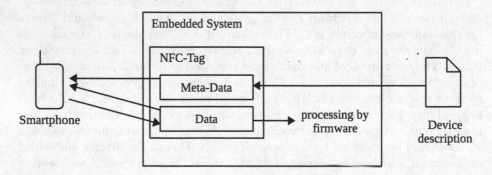

Fig. 2. Configuration of an embedded device with NFC

The components of the system (embedded device, smartphone, and artifacts for firmware development) and their interactions are shown in Fig. 3.

When writing the firmware, the embedded developer specifies the device properties and available settings in a simple YAML-based text document, according to format proposed in Sect. 5. A provided utility detailed in Sect. 6 translates that human-readable description into two binary files consisting of data

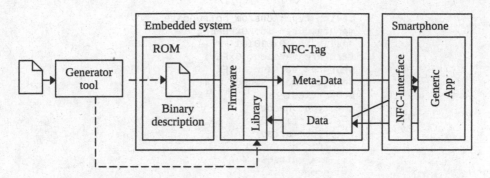

Fig. 3. System architecture for embedded device configuration

with default values and compressed metadata. Both can be combined as a single blob for storage in a backup ROM or as separate blobs for initialization of the NFC chip. Note that the metadata does not need to be understood by the device firmware. In addition, the utility serves as a code generator that generates a C struct and accompanying deserialization and serialization functions, which provide firmware developers with natural access to the stored configuration values. The specification of the device settings is close to existing modern C data types, to simplify their use for embedded developers and to allow easy mapping of settings values to generated struct members. To this end, type sizes are specified explicitly and strings are fixed length and zero terminated. For dates and times, custom structs and types are provided in the generated code, which follow the conventions of the modern C++ date and time library. Furthermore, code to perform sanity checks on the deserialized values is generated and provided as C functions for the firmware developer to use. The embedded developers only need to concern themselves with the specifics of the used hardware, which means implementing functions that read/write binary data from/to the NFC chip and implement reactions to interrupts triggered by the NFC hardware. Typically, the entire information stored on the NFC tag is less than a kilobyte and thus all input/output with the NFC tag is completed in a single operation. Optionally, the metadata can specify that only the data needs to be written back by the smartphone. In this case, the firmware developer has to restore the metadata after every write operation to the NFC tag. However, the benefit is a faster completion of the transfer from smartphone to device, as the data portion is typically small compared to the metadata. This means the smartphone needs to be in contact with the NFC tag for a much shorter time.

5 Device Specification of Configuration Data

To specify how a device is configured, the device developer creates a description of its features and available settings. This description uses the YAML format with a custom extension. An example can be seen in Fig. 4. Here, a time controlled power socket is defined, with four outlets which can be turned on or off during a

```
1   device_type: Custom_NoTruncate
2   manufacturer_id: 0xFEFEFE
3   device_id: 0x010101
4   firmware_version: 0xFE
5   protocol_version: 0xFE
6   device_name: "socket timer"
7   properties:
8     - type: header
9       id: "title"
10    - type: array
11      id: "program"
12      max_entries: 4
13      properties:
14        - type: bool
15          id: "program_active"
16          dependencies:
17            True:
18              - "active_range"
19              - "active_sockets"
20            False:
21          default: True
22        - type: divider
23        - type: time_range
24          id: "active_range"
25          default: "00:00:00;23:59:59"
26        - type: n_out_of_m
27          id: "active_sockets"
28          entries:
29            - "sock1"
30            - "sock2"
31            - "sock3"
32            - "sock4"
33  translation_data:
34    #include "en.yaml"
```

Fig. 4. Device specification for a timed power socket

specified time range. Every specification starts with a block of device information (lines 1–6). It contains unique identifiers for the device and information about its type. This is followed by a list of all available settings (lines 7–32). Each setting has a type and, dependent on the type, zero or more additional properties. A setting of type bool for example contains an identifier, the identifiers of settings which depend on its state, and a default initialization value. The type array is special, as additional settings can be nested within it. All nested settings will be repeated a given number of times (max_entries). This allows storing multiple sets of the same data, while not inflating the metadata description with repeated setting declarations. At the end of the declaration, a list of translations is defined (lines 33–34). Here, the custom addition to the YAML format can be seen in use. An include system, analogous to the C preprocessor include mechanics, can be used to extract parts of a specification into separate files for easier reuse. As YAML is indent-sensitive, the indentation of the #include statement is applied to all included lines when expanding the statement. The include statements are processed by the provided tool chain before parsing the actual YAML content. As the character # begins a comment in YAML, the documents are still valid YAML without preprocessing, albeit without the included content.

```
1   - language: "en"
2     translations:
3       "title": "Time controlled outlet"
4       "program": "Programs"
5       "program_active": "Program enabled"
6       "active_range": "Active time range"
7       "active_sockets": "Active outlets"
8       "sock1": "Socket 1"
9       "sock2": "Socket 2"
10      "sock3": "Socket 3"
11      "sock4": "Socket 4"
```

Fig. 5. Translation specification

Translations are used to display the names of settings in different languages in the smartphone application. They are defined as a list of entries, each defining a language and a mapping as pairs of identifiers to translate and their corresponding translations, as seen in Fig. 5.

6   Tool Chain

To help developers with integration of their devices into the system, a custom tool chain is provided. It is implemented in Python and handles code and *blob* generation. It can either be invoked directly from the command line, or imported into other scripts. As seen in Fig. 6, the package takes a device specification as input and generates a corresponding binary blob and C code.

The blob contains the metadata and data in a packed binary format. The metadata is additionally compressed per default, although this can be disabled. This blob is typically not stored directly on the NFC tag. Rather, it should be stored in non-volatile memory on the device, where the embedded system reads the metadata and data blocks and transfers them to the tag during initialization. This allows for an easy factory reset. If the metadata contains translations, developers can optionally specify which languages are packed into a metadata block, which enables regional customization. The blob can then contain several metadata blocks where each contains one of the specified language sets. This is useful, if the NFC tag is too small to hold all languages at once. The device can then load the metadata block into tag memory, which contains only the requested language. To make reading of the generated blob easier for the embedded developer, a table of contents is added to the beginning of the blob. It contains start index, length, and a checksum or hash of each contained block. The checksum can be used to detect changes to or corruption of the blocks on the NFC tag.

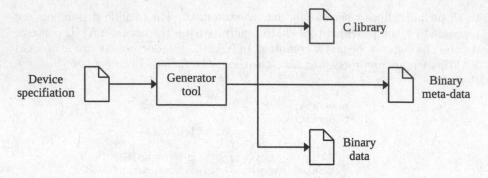

Fig. 6. Artifacts generated by the toolchain

Currently, *MD5*, *SHA1*, *SHA256*, and *CRC32* are available, to allow the developer to choose the most convenient checker available on the embedded platform. In addition to the binary blob, C source code is generated. This code contains a struct definition, conversion functions, and general helper functions. The struct contains members for every setting declared in the device specification. With the generated conversion functions, the values in a binary data block can be transferred into an instance of the aforementioned struct or vice versa. Conversion from struct to binary may be used by the firmware to change configuration values in the NFC storage. Optionally, functions can be generated to perform sanity checks on configuration values and ensure the use of valid values. For an embedded developer, the only steps necessary are creating a device description, generating the blob and code and then implementing writing the blob to an NFC tag, reading the data, and using the generated code to fill a struct with the read data. As a result, the firmware receives a ready-to-use struct, containing only commonly used C data types.

7   Generic Configuration App

We have implemented a reference implementation of the proposed configuration protocol with a generic Android app. This app allows to read a configuration from a device, modify it and then write it back to the device. Additionally, users can save, recall, or share modified configurations. The app supports configuring any device adhering to the aforementioned device descriptions. Note, that this is achieved without needing any kind of device specific code adjustments on the app side. Touching a tag triggers reading the configuration from a device. The data on the tag contains a URL, which is associated with the Android app. This causes launching the app if it is installed, or otherwise opens the URL in a browser, which may offer a download. The application receives the contents of the NFC tag from the Android system and uses a Java library to convert the contained metadata and data into an internal representation of the device configuration. This functionality is extracted into a separate library to ease reuse in other

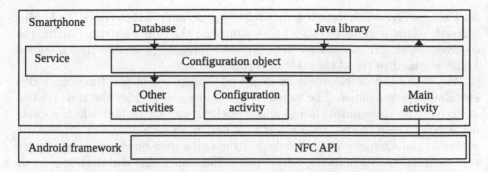

Fig. 7. Architecture of the Android application

applications. This representation is handed to an activity, which constructs a dynamic user interface from it, displaying the available settings and their values and allowing modification. The flow of the tag's data through the application components during this process is shown in Fig. 7.

To create this dynamic interface, we use a library of self-contained widget components, that each handles display and modification of one type of settings. There are some types where a value can be set in different ways. These have multiple interaction components, each implementing a different way to display and input values. As an example, numbers can be input as text or through setting a slider to the desired position. For some settings, such as water temperature, writing a number as text is fine, but for others, such as the brightness of a lamp, a slider might be more comfortable to use. Users may choose their preferred interaction style for a configuration setting per device permanently. The default interaction style is based on heuristics that may change over time and explicitly not embedded into the metadata, as these increase the metadata and only express personal preferences of developers and not necessarily users.

After a user has changed the configuration values, these can be written back to the device. In addition, a copy may be stored in the application for later reuse. This entails, that previously saved configurations can be loaded, changed, deleted, stored and applied. These saved configurations can also be exported as a file in order to share with other users. Importing these files is also possible. In the future, optional sharing over a website can be envisioned.

8   Evaluation

A Raspberry Pi is used to simulate different embedded devices. To this end, it is connected to a custom board hosting an NFC tag and several LEDs.

With this setup, a series of fictional devices are simulated. Based on a specification the generated code is tested, and each simulated device is configured by the generic smartphone application. To simulate a device, the generated binary metadata and data is written on the NFC tag. As soon as a smartphone writes back the configuration data, a command line program reads the data block, converts the data and fills the struct after applying the sanity checks. Finally, the received values are printed on the command line. If applicable, the embodied LEDs are used to signal the state of certain settings.

We detail the development process and simulation of the time controlled switchable power outlet. The values to configure are whether the time control is active, the time range during which sockets are active and which sockets are active. These values are grouped together in a profile and four profiles are available and configurable on the device. First, the developer creates the device specification, as seen in listing 4, from which the binary blob and the library code is generated. For the simulation we use the generic driver to read and write from the development board containing the NFC tag. After touching the tag with the smartphone the app opens, the metadata and data is read and presented to the user as shown in Fig. 8a). The user can modify the setting values, save them in the application as in Fig. 8b), or load previously stored values as in Fig. 8c).

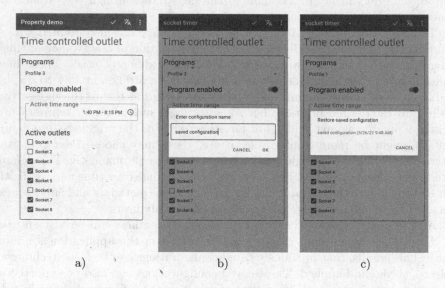

Fig. 8. Android application user interface for configuration

Accepting modified values triggers a transition to the transmission screen, where the user may request to write the configuration back onto the device. As soon as the smartphone then recognizes the NFC tag again, the changed configuration gets transmitted. Every time a new device is configured, the application stores its identification and associated user customization preferences in a database. These saved preferences can be accessed, modified, or deleted in the application's settings menu, which is shown in Fig. 9a). This is also, where saved configuration values can be managed. In this menu shown in Fig. 9b), saved configurations can be imported, exported or deleted.

Fig. 9. Android application user interface for management of saved configurations

The presented architecture and implementation is sound and has been used regularly without issues. Binary blob and source code generation is reliable and quick. Because of the generic approach to the smartphone application, a vast range of devices can be supported without further modification. Issues experienced during the realization mostly came down to the used NFC hardware. The NFC tag used in the simulator (NXP NTAG I^{2}C *plus* [13]) reaches only very low write speeds of approximately 400 bytes per second. Because of this, applying a configuration may take up to two seconds. This means, the smartphone needs to be near the tag for a long time, which makes writing susceptible to transmission errors. To combat this, the data format has been optimized for low memory requirements. Data is packed as efficiently as possible and optional metadata compression leads to a size reduction of up to 60%. Compression rates for different device examples are collected in Table 2.

Table 2. Metadata compression efficiency

Device	Size (bytes)	Compressed (bytes)	Ratio
Demo	1060	592	55,8%
Demo with translations	3451	1324	38,4%
Heater	295	179	60,7%
Heater with translations	888	449	50,6%

The option to only write back data and leave out the metadata shortens transmission time considerably, but requires that the firmware developer restores the tag metadata as well as the data after each configuration. Tag memory is also a scarce resource, with common tags having at most about one kilobyte capacity. With the presented size reduction options, this is usually no hindrance, except for devices with unusually complex configurations. As the size of binary blobs is fixed, a developer can choose a tag with appropriate capacity. Accidentally or maliciously overwriting tags with invalid data is also possible. This can only be solved by the embedded developer, as tag protection mechanisms are tag specific. However, the generated checksums help with detection of data corruption, while resetting needs to be implemented by the embedded developer.

The Android application works reliably and implementation is straightforward. Minor performance issues are the result of the dynamic creation of the configuration screen. If many widgets are created, which needs to happen in the main thread after the tag is read, delays might become noticeable for the user. However low hundreds of milliseconds were only recorded on a kitchen sink application using all widget types on a low-end smartphone.

9   Conclusion

With the proposed approach we allow easy configuration of hardware-constrained disconnected devices with a smartphone without internet connection. To that end, we can use any suitable NFC chip on the device, a generic app for any NFC-enabled Android smartphone, and a tool chain that translates device configuration specifications to commonly used C data structures for the embedded device developer. Thus, we cater to the needs of privacy-aware consumers and business customers and in addition, increase sustainability of the devices by freeing them from proprietary vendor lock-ins. Obviously, the embedded device vendor has to support the effort. However, by giving up on the necessity (not the option) for cloud connectivity, embedded device developers do not need to run remote services and can save on hardware costs, while still being able to give users an easy-to-use, universally compatible configuration system. The tool chain has been demonstrated to work on lab prototypes using both embedded devices and a Raspberry Pi test bed. The tool chain and the app have been released as open source under https://github.com/ni9l/eput-tools/. Next steps include pilot projects with embedded devices and collecting feedback from the embedded developers. Obviously, newer versions of the app have to support all existing devices configured with earlier tool chain versions. Thus, versioning of the specification format has to be established. When successfully used in more and more embedded devices along with further adoption a standardization process can be envisioned. We hope that this project helps to demonstrate that ease of use, low hardware cost, privacy, and sustainability are attainable simultaneously.

References

1. Abukwaik, H., Groß, C., Aleksy, M.: NFC-based commissioning of adaptive sensing applications for the 5G IIoT. In: Barolli, L., Hellinckx, P., Enokido, T. (eds.) BWCCA 2019. LNNS, vol. 97, pp. 150–161. Springer, Cham (2020). https://doi.org/10.1007/978-3-030-33506-9_14
2. Almeida, N., Alemany, R., Glege, F., da Silva, J., Varela, J.: A software package for the configuration of hardware devices following a generic model. Comput. Phys. Commun. **163**(1), 41–52 (2004)

3. Argueta, D., et al.: Enhancing the restaurant dining experience with an NFC-enabled mobile user interface. In: Memmi, G., Blanke, U. (eds.) MobiCASE 2013. LNICST, vol. 130, pp. 314–321. Springer, Cham (2014). https://doi.org/10.1007/978-3-319-05452-0_29

4. Bisignano, M., Di Modica, G., Tomarchio, O.: An "intent-oriented" approach for multi-device user interface design. In: 20th International Conference on Advanced Information Networking and Applications - Volume 1 (AINA 2006), vol. 2, pp. 186–194 (2006)

5. Eisenmenger, W., Meßmer, J., Wenger, M., Zoitl, A.: Increasing control application reusability through generic device configuration model. In: 2017 22nd IEEE International Conference on Emerging Technologies and Factory Automation (ETFA), pp. 1–8 (2017)

6. How we built facebook lite for every android phone and network. engineering. fb.com/2016/03/09/android/how-we-built-facebook-lite-for-every-android-phone-and-network/. Accessed 21 May 2022

7. Haase, J., Meyer, D., Eckert, M., Klauer, B.: Wireless sensor/actuator device configuration by NFC. In: 2016 IEEE International Conference on Industrial Technology (ICIT), pp. 1336–1340 (2016)

8. Homann, M., Banova, V., Oelbermann, P., Wittges, H., Krcmar, H.: Towards user interface components for dashboard applications on smartphones. In: Memmi, G., Blanke, U. (eds.) MobiCASE 2013. LNICST, vol. 130, pp. 19–32. Springer, Cham (2014). https://doi.org/10.1007/978-3-319-05452-0_2

9. Kim, W., Ko, H., Yun, H., Sung, J., Kim, S., Nam, J.: A generic internet of things (IoT) platform supporting plug-and-play device management based on the semantic web. J. Ambient Intell. Human. Comput. (2019). https://doi.org/10.1007/s12652-019-01464-2. ISSN 1868-5145

10. Lotito, A., Mazzocchi, D.: OPEN-NPP: an open source library to enable P2P over NFC. In: 2012 4th International Workshop on Near Field Communication, pp. 57–62 (2012)

11. Matos, A., Romão, D., Trezentos, P.: Secure hotspot authentication through a near field communication side-channel. In: 2012 IEEE 8th International Conference on Wireless and Mobile Computing, Networking and Communications (WiMob), pp. 807–814 (2012)

12. NFC Forum: NFC Data Exchange Format (NDEF). NFC Forum, 1.0 edn. (2006)

13. NXP Semiconductors: NTAG I^{2}C plus: NFC Forum T2T with I^{2}C interface, password protection and energy harvesting. NXP Semiconductors, 3.5 edn. (2019)

14. Schüpbach, A., Baumann, A., Roscoe, T., Peter, S.: A declarative language approach to device configuration. ACM Trans. Comput. Syst. **30**(1), 1–35 (2012)

15. The European Parliament and the council of the European Union: Regulation (EU) 2016/679 of the European parliament and of the council of 27 April 2016 on the protection of natural persons with regard to the processing of personal data and on the free movement of such data, and repealing directive 95/46/EC (general data protection regulation) (2016). data.europa.eu/eli/reg/2016/679/oj

16. Ulz, T., Pieber, T., Höller, A., Haas, S., Steger, C.: Secured and easy-to-use NFC-based device configuration for the internet of things. IEEE J. Radio Freq. Identif. **1**(1), 75–84 (2017)

17. USB Implementers' Forum: Device Class Definition for Human Interface Devices (HID). USB Implementers' Forum, 1.11 edn. (2001)

18. Defined class codes. www.usb.org/defined-class-codes. Accessed 12 May 2022

19. Škrinjar Vidović, M.: EU data protection reform: challenges for cloud computing. Croatian Yearb. Eur. Law Policy **12**(1), 171–206 (2016). hrcak.srce.hr/174312

Decentralized Federated Learning Methods for Reducing Communication Cost and Energy Consumption in UAV Networks

Deng Pan[1]([⊠]), Mohammad Ali Khoshkholghi[2], and Toktam Mahmoodi[2]

[1] University College London, London, UK
deng.pan.22@ucl.ac.uk
[2] King's College London, London, UK
{ali.khoshkholghi,toktam.mahmoodi}@kcl.ac.uk

Abstract. Unmanned aerial vehicles (UAV) or drones play many roles in a modern smart city such as the delivery of goods, mapping real-time road traffic and monitoring pollution. The ability of drones to perform these functions often requires the support of machine learning technology. However, traditional machine learning models for drones encounter data privacy problems, communication costs and energy limitations. Federated Learning, an emerging distributed machine learning approach, is an excellent solution to address these issues. Federated learning (FL) allows drones to train local models without transmitting raw data. However, existing FL requires a central server to aggregate the trained model parameters of the UAV. A failure of the central server can significantly impact the overall training. In this paper, we propose two aggregation methods: Commutative FL and Alternate FL, based on the existing architecture of decentralised Federated Learning for UAV Networks (DFL-UN) by adding a unique aggregation method of decentralised FL. Those two methods can effectively control energy consumption and communication cost by controlling the number of local training epochs, local communication, and global communication. The simulation results of the proposed training methods are also presented to verify the feasibility and efficiency of the architecture compared with two benchmark methods (e.g. standard machine learning training and standard single aggregation server training). The simulation results show that the proposed methods outperform the benchmark methods in terms of operational stability, energy consumption and communication cost.

Keywords: Federated Learning · Unmanned Aerial vehicles · Decentralized Training

1 Introduction

Unmanned aerial vehicles (or drones) will positively impact society by supporting various services for the modern smart city. Examples include applications in

J. Taheri et al. (Eds.): MobiCASE 2022, LNICST 495, pp. 18–30, 2023.
https://doi.org/10.1007/978-3-031-31891-7_2

goods delivery, real-time road traffic monitoring, target identification, etc. [1]. Machine learning (ML) to give UAVs network intelligence is a crucial requirement to enable such applications. However, traditional machine learning techniques require uploading all data to a cloud-based server for training and processing, which represents a considerable challenge for drone swarms [2].

In a first consideration, the data generated by drones may be sensitive, and could be intercepted while uploading the data to the cloud, leading to a privacy breach. Secondly, drones' large numbers of data can result in impractical delays when uploading, thus creating a time lag for swarms of drones that prevents them from conducting real-time monitoring. Finally, drones can consume a great amount of energy when training models, meaning there may be related challenges to doing so in terms of energy constraints [3].

Distributed machine learning techniques represent a new solution to address these issues and challenges, whereby drones train machine learning models without sharing raw data. Federated learning (FL), recently developed and proposed by Google as an emerging distributed machine learning technology, will provide further new technology to support the intelligence of drones [4]. The concept of federated learning is allowing each drone to train its learning model based on its data. The parameters of each drone's trained model are then sent to a parameter server to update the model for a new round of training, without sending the raw data to the cloud. This training model allows for reasonable data security, latency and energy consumption.

However, the highly mobile nature of drones means conventional FL is not well-suited, given their complex working environment. If the parameter server does not work properly, it will impact the training effectiveness of the whole UAV network [5]. At the same time, each drone needs to return to the parameter server after each local training session to upload parameter information; and then return to where it is supposed to work after updating the parameters, which increases the working time and wear and tear on the drone's components. Also, if a large number of UAVs access the parameter server simultaneously to upload parameter information and update the local model, it is a test of the bandwidth of the parameter server [6]. In this context, a fully decentralised federated learning architecture holds the potential to significantly optimise existing intelligent drone networks.

To address the current problem, Decentralized Federated Learning for UAV Networks (DFL-UN), proposed by Qu et al., proposes establishing links between UAVs in a small area while setting any UAV in range to act as a simulation server to aggregate models [7]. The authors address FL's centralisation and user selection issues for UAV network but do not account for UAVs' limited power supply. The main objective of this work is to optimise the architecture of the DFL-UN while guaranteeing training results and analysing communication cost and energy consumption. As our contributions in this work, first, classify UAVs in the region into two groups to determine which two UAVs will be used as aggregation centres. Then, we optimise the decentralised algorithm proposed by Liu et al. to derive two aggregation methods, Commutative FL and Alternate FL

[8]. The method we proposed is compared with cases where FL is not used, with only one aggregation centre. The simulation results validate the convergence of commutative FL and alternate FL in the architecture, demonstrating more stable training results than when not using FL and with only one aggregation centre.

The rest of this paper is outlined below. We will briefly introduce the relevant research literature in Sect. 2. Then we will briefly describe the design and implementation of this work in Sect. 3. Nest, we will discuss the collected and listed research findings in Sect. 4. The conclusion will be given in the final Sect. 5.

2   Background and Related Work

2.1   Definition of Federated Learning

The term federated learning (FL) was first coined by McMahan et al. in 2016, who defined it as "a learning task that is solved by a loosely federated participating device (client) that is coordinated by a central server" [4]. We can now define FL more broadly by adopting this initial definition: FL is a privacy-preserving distributed machine learning technique where N participants (clients) train the same model using their locally stored data. Eventually, producing a federated model on a central server by exchanging and aggregating the model parameters for each client and using the federated model to update the model on the client's side [9,10]. The framework of FL is shown in Fig. 1.

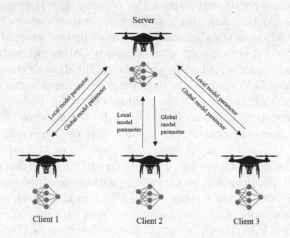

Fig. 1. FL framework

2.2   Overview of the Federated Average Algorithm (FedAvg)

The literature [4] proposes using the federated averaging algorithm (FedAvg) to train models in federated learning. Assuming that the client has the initial model, in round t when the central model parameters are updated, the k-th participant will compute the batch gradient g_k, and the server will aggregate these gradients and use the updated information on the model parameters according to the following formula.

$$w_{t+1} \leftarrow w_t - \eta \sum_{k=1}^{K} \frac{n_k}{n} g_k \tag{2-3}$$

where η is the learning rate and $\sum_{k=1}^{K} \frac{n_k}{n} g_k = \Delta f(w_t)$. FedAvg is the average gradient w sent to each participant, who will calculate the updated model parameters w according to Eq. 2-3.

$$\forall k, w_{t+1}^k \leftarrow w_t - \eta g_k \tag{2-4}$$

$$w_{t+1} \leftarrow \sum_{k=1}^{K} \frac{n_k}{n} w_{t+1}^k \tag{2-5}$$

Each client performs gradient descent locally on the existing model parameters w_t using local data according to Eq. 2-4 and sends the locally updated model parameters $w_{t+1}^{(k)}$ to the server. The server then computes a weighted average of the model results according to Eq. 2-5 and sends the aggregated model parameters w_{t+1} to each participant. The complete pseudo-code of FedAvg is as follows.

2.3   Decentralised Federated Learning for UAV Networks

As stated in the literature [7], FL training may be aborted due to the mobility of UAV. DFL-UN represents a solution to use in such situations. The prominent architecture of DFL-UN is based on a fully distributed scenario without a central server or a fixed UAV as a parameter server. Each drone is trained with local data, and neighbouring drones receive the model parameters. The DFL-UN architecture operates in the following steps.

- All drones are pre-installed with the FL training model. A built-in coordinator is responsible for distributing central information to all designed drones and monitoring the FL training process.
- Each drone will train a local model parameter in training round t.
- Firstly, in round t, drones $i + 1$, $i + 2$ and $i + 3$ send their training models to $W_{i+1,t}$, $W_{i+2,t}$ and $W_{i+3,t}$ to drone i, where $W_{i,t}$ represents drone i's local model parameter at training round t.
- Secondly, drone i aggregates $W_{i+1,t}$, $W_{i+2,t}$, $W_{i+3,t}$ and including drone i's training model parameter to generate an aggregated model parameter.

- Thirdly, drone i will "broadcast" these aggregated model parameters to its neighbouring drones for model updates, and drone i will also update its local model.

Algorithm 1. FederatedAveraging. K clients are indexed by k; C is the percentage of clients performing the computation in each round, B is the local minibatch size, E is the number of local epochs, $\mathcal{P}_k$ denotes the index set located on the data side of participant k and η is the learning rate [4]

Server executes:

1: initialize w_0
2: **for** each round $t = 1, 2, \ldots$ **do**
3:　　$m \leftarrow \max(C \cdot K, 1)$
4:　　$S_t \leftarrow$ (random set of m clients)
5:　　**for** each client $k \in S_t$ **in parallel do**
6:　　　　$w_{t+1}^k \leftarrow$ ClientUpdate (k, w_t)
7:　　**end for**
8:　　$w_{t+1} \leftarrow \sum_{k=1}^{K} \frac{n_k}{n} w_{t+1}^k$
9: **end for**

ClientUpdate(k, w): //Run on client k

1: $\mathcal{B} \leftarrow$ (split $\mathcal{P}_k$ into batches of size B)
2: **for** each local epoch i from 1 to E **do**
3:　　**for** each batch $b \in \mathcal{B}$ **do**
4:　　　　$w \leftarrow w - \eta \nabla \ell(w; b)$
5:　　**end for**
6: **end for**

2.4　Decentralised Federated Learning (DFL)

An algorithm that alternates between local updates and inter-node communications is proposed in the literature [8] to cope with the absence of a central server in DFL. This algorithm is broadly similar to FedAvg but takes a new approach to decentralisation. Liu et al. proposed dividing the communication steps into τ_1 and τ_2, i.e. the frequency of local updates and the frequency of inter-node communications, respectively [8]. Let $\tau = \tau_1 + \tau_2$ be defined in the DFL framework as an iteration round, such that for the kth iteration round, $((k-1)\tau, k\tau)$ for k = 1, 2, ... The complete pseudo-code is as follows.

Algorithm 2. DFL[8]

Parameters:

　　Learning rate η, total number of steps T, computation frequency τ_1 and communication frequency τ_2 in an iteration round, $C \in \mathbb{R}^{N \times N}$ is the confusion matrix and X_t is the model parameter matrix

1: Set the initial value of X_0
2: **for** $t = 1, 2, \ldots, K_\tau$ **do**
3:　　**if** $t \in [k]_1$ where $k = 1, 2, \ldots, K$ **then**
4:　　　　$X_{t+1} = X_t - \eta G_t$　　*local update*
5:　　**else**
6:　　　　$X_{t+1} = X_t C$　　*inter − node communication*
7:　　**end if**
8: **end for**
9: **for** $t = K_{\tau+1}, \ldots, T$ **do**
10:　　$X_{t+1} = X_t - \eta G_t$
11: **end for**

---

3 Design and Implementation

3.1 UAV Model

In this work, 10 or 20 drones were randomly generated in a 10×10 m^2 area. The drones are then divided into two clusters using the K-means clustering algorithm, and the coordinate information of each drone is recorded by themselves [11]. The default drone had enough battery to support the drone for 1 h of flight and operation. Next, each drone was assigned a model and local data to be used for training. Finally, the coordinates, client ID and remaining battery level (first recorded as 100%) of all drones in the current training round were recorded for subsequent data analysis.

3.2 Aggregation Model

For training that required FL, we determined whether the cluster head was single or double, so the appropriate training function could be selected. The training functions take different patterns depending on the training method, of which there were four, based on different UAV classifications, in this work:

- Commutative FL (C). This model is the case of classifying the UAV into two clusters, where the client and aggregate server within each cluster perform n intra-cluster FL and m inter-cluster exchanges of FL.
- Alternate FL (A). This model is the case where the drone is divided into two clusters, and the subsequent training round after each intra-cluster FL will be the inter-cluster swapped FL.
- One-server FL (One). This is one of the control group. In this model, the UAV is grouped into one cluster, similar to the FL in the standard case.

– Normal Machine Learning (O). This is another one of the control group. In this model, each drone will only be trained with local data, and no data exchange occurs.

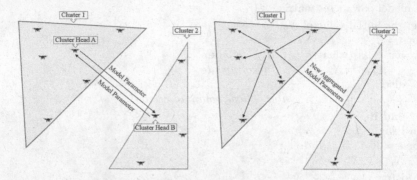

Fig. 2. Inter-cluster FL framework

The inter-cluster FL exchanged between the clusters mentioned above refers to an additional round of aggregation and evaluation computation by two clusters uploading each other's updated model parameters after the FL ends within each the cluster. Suppose the two cluster heads are A and B, and aggregated model performance sent by B as the client to A as the server is better than that sent by A as the client to B as the server. All drones in both clusters will receive the new aggregated model parameters sent by B to A and vice versa. The framework of inter-cluster FL is shown in Fig. 2. The process of training and updating the drone information is repeated until a valid model with stable accuracy and loss values is obtained. After this, the accuracy, loss, battery remaining, data sent, data received and total data sent and received for each round were evaluated.

3.3   Energy Consumption Model

Local Computation. The drone will perform a deep learning strategy for image recognition locally, and this work calls on the GPU to train it while performing the simulation. For this, we will fix the energy stored in each drone battery as E_d and then record the average power of the GPU during the training call to the GPU, P_{avg}. The training time, t_{tr} for each drone will then be recorded, and the estimated computational energy consumption will be calculated using $E_c = P_{avg}t_{tr}$. It was eventually converted to percentage units by Batter used $= E_c/E_d$.

Communication Energy. In this work, all drones are assumed that the wireless channel between each other is connected by Line of Sight (LOS) wireless link. Similar to [12], the communication between drone k and another drone is represented as follows:

$$E_k^C = t_k p_k$$

where t_k is the time duration to transmit data of size s and p_k is the average transmit power of drone k. In this work, t_k will be calculated according to the theoretical minimum time, which is represented as follows:

$$t_k^{min} = \frac{s}{b_k \log_2 \left(1 + \frac{g_k p_k}{N_0 b_k}\right)}$$

where b_k is the bandwidth allocated to drone k, g_k is the channel gain between two drones and N_0 is the power spectral density of the Gaussian noise. As in [13], we denote the coordinates of UAV k and UAV $k+1$ as $q = \{x, y, z\}$ and $q' = \{x', y', z'\}$, respectively. The channel gain g_k of drone k can be calculated by $g_k = \hat{\beta}_0 \left(d_k/d_0\right)^{-\alpha}$, where $\hat{\beta}_0$ is the reference channel gain at $d_0 = 1$ m, d_k is the distance between the two drones, and α is the path loss parameter. Hence, the 3D-coordinate distance d_k is given as $d_k = \sqrt{||q - q'||}$. It is important to note the size of t_k^{min} is mainly determined by the distance between the drones and the amount of transmitted data.

3.4   Communication Cost Model

In this work, communication costs will be recorded into an logbook file by recording each drone's sent and received file size and the sum of received and sent data. Also, for each method, the communication cost of the drones will be averaged over the file size of all drones received and sent and the sum of received and sent.

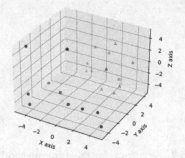

Fig. 3. 20 Drones UAV model: Blue indicates cluster 1, yellow indicates cluster 2 and the red indicate the cluster head (Color figure online)

4   Simulation Result

In this section, we evaluate the performance of the Commutative FL and Alternate FL and verify the validity of these methods by showing the numerical results of the simulations. The simulated UAV model is first built according to the UAV model in Sect. 3, as shown in Fig. 3. The datasets used by the UAV for the machine learning model were all taken from the coloured image dataset

Cifar-10 [14]. By default, each drone was loaded with a local database of 5000 pieces of information. In addition, the machine learning model used for the UAV uses the deep convolutional network model ResNet-18 [15]. We set the energy possessed by each UAV battery to $E_d = 274$ Wh. The bandwidth of each UAV is $b_k = 20$ MHz. The reference channel gain is set to $\hat{\beta}_0 = 28 + 20\log_{10}(f_c)$ according to the reference path loss specified in [16], where the carrier frequency $f_c = 2$ GHz. The path loss exponent is set to 2.2, the noise power spectral density $N_0 = -174$ dBm/Hz and the average transmitted power $p_k = 10$ dB. For the parameters mentioned above, if not specifically labelled, most parameters are referenced in [12,13]. After this, to avoid ambiguity, we call the completion of a training ground by the drone a local epoch (le) and the completion of a federal learning training round a global epoch (ge). The number of federation learning training rounds performed by two clusters in this cluster is called local round (lr), and the number of interactions between the two clusters is called global round (gr). If not specifically labelled, the default ge will be 30.

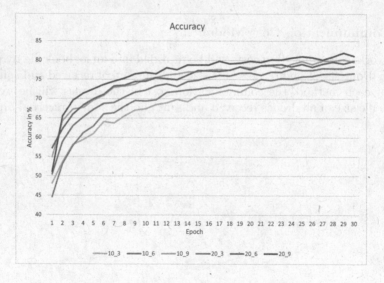

Fig. 4. Accuracy for training method C

Figures 4 shows the model's accuracy for training method C, when lr is 5 and gr is 10, and for le values of 3, 6 and 9, respectively. It can be seen from the graphs that 20 drones are generally more accurate than 10 drones when at an equal le. This phenomenon was predictable because 20 drones carry twice the amount of data as 10 drones, and therefore, the model should generate this more accurate prediction. We also found that higher le values tended to bring about higher accuracy with the same number of drones. Again, this phenomenon was predictable as more le means faster convergence and more training rounds. At the same time, the accuracy of the two clusters was very close, indicating that the architecture was stable for training method C. As shown in Table 1, our simulations show that for training method C, different combinations of lr and gr do not have a clear impact on the accuracy of the training but only on the energy consumption and the data amounts sent and received.

Table 1. Data relating to training method C

Type	Accuracy	Loss	Avg. Battery	Avg. Send/GB	Avg. Receive/GB	Avg. S&R/GB
C_5lr_5gr_10	75.33	0.7242	92.00	2.47	2.47	4.94
C_5lr_5gr_20	76.68	0.6875	88.04	3.23	3.23	6.47
C_5lr_15gr_10	75.05	0.7201	93.45	2.67	2.67	5.34
C_5lr_15gr_20	77.24	0.6642	92.49	2.90	2.90	5.79
C_5lr_10gr_10	74.98	0.7272	92:27	2.67	2.67	5.34
C_5lr_10gr_20	76.77	0.6749	92.46	2.90	2.90	5.79
C_15lr_5gr_10	74.96	0.7293	93.07	2.07	2.07	4.15
C_15lr_5gr_20	76.78	0.6793	93.11	2.31	2.31	4.61
C_10lr_5gr_10	75.18	0.7146	93.85	2.27	2.27	4.54
C_10lr_5gr_20	77.09	0.6743	93.15	2.50	2.50	5.01

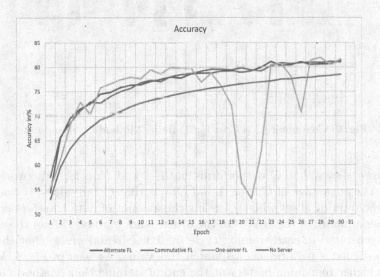

Fig. 5. Accuracy for training method C, A, One and O

Next, we compare training methods C, A and One. In the next comparison, training method C will fix the value of lr to 5 and the value of gr to 5. Figure 5 shows how the proposed method compares with other traditional methods in the case of 20 drones. It can be seen that the two proposed methods outperform traditional machine learning methods in terms of accuracy and stability over a single drone federation learning network. The stability of method One with a number of 10 drones is comparable to methods C and A. However, it has been tested that for method One, the training results fluctuate significantly when the number of drones exceeds 15 in a network. This phenomenon suggests that in the case of method *One* if the number of drones in the network exceeds 15, the overlap in the data allocated by the drones is higher. This can lead to an overfitted training model, resulting in unstable test accuracy.

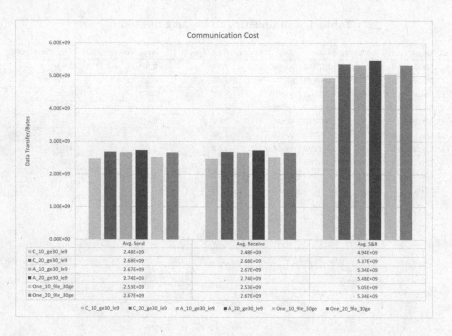

Fig. 6. Communication Cost for Training Method C, A and *One*

Figure 6 shows that C outperforms both A and *One* for a network of 10 drones. The advantage of C over A and *One* is that C can adjust lr and gr to ensure the training effect while adjusting the amount of data transfer according to the performance of the UAV. On the other hand, Method A fixes the amount of data generated in each training cycle by fixing lr and gr, so that the UAV can be assigned a performance to match this value. As seen in Fig. 7, Method C has a higher remaining battery at the end of training than Method *One* and Method A. Methods C, and A are very similar in terms of the amount of battery remaining at 10 and 20 UAVs. On the other hand, Method *One* has a much lower battery left than Methods C and A due to the much higher amount of data handled by a single drone as a parameter server, especially at 20 drones. Again, those figures show the proposed method outperforms conventional FL and ML regarding communication consumption, battery left and training stability.

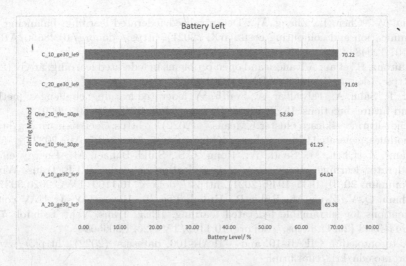

Fig. 7. Battery Left for Training Method C, A and *One*

5   Conclusion

In this paper, we investigate the problem of training methods for decentralised federated learning UAV networks. We optimise and propose two learning methods based on existing decentralised federated learning networks to cope with UAVs' communication cost and energy consumption. Simulated numerical results show that our proposed new learning methods can effectively guarantee training results while outperforming conventional training methods in terms of the training stability, communication cost and energy consumption.

References

1. Shakhatreh, H., et al.: Unmanned aerial vehicles (UAVs): a survey on civil applications and key research challenges. IEEE Access **7**, 48572–48634 (2019)
2. Abdulrahman, S., Tout, H., Ould-Slimane, H., Mourad, A., Talhi, C., Guizani, M.: A survey on federated learning: the journey from centralized to distributed on-site learning and beyond. IEEE Internet Things J. **8**(7), 5476–5497 (2021)
3. Brik, B., Ksentini, A., Bouaziz, M.: Federated learning for UAVs-enabled wireless networks: use cases, challenges, and open problems. IEEE Access **8**, 53841–53849 (2020)
4. McMahan, H.B., et al.: Communication-efficient learning of deep networks from decentralized data. arXiv (2016). https://doi.org/10.48550/ARXIV.1602.05629
5. Bonawitz, K., Eichner, H., Grieskamp, W., Huba, D., Ingerman, A., Ivanov, V., et al.: Towards federated learning at scale: system design (2019)
6. Elgabli, A., Park, J., Bedi, A. S., Bennis, M., Aggarwal, V.: GADMM: fast and communication efficient framework for distributed machine learning (2019)
7. Qu, Y., et al.: Decentralized federated learning for UAV networks: architecture, challenges, and opportunities. arXiv (2021). https://doi.org/10.48550/ARXIV.2104.07557

8. Liu, W., Chen, L., Zhang, W.: Decentralized federated learning: balancing communication and computing costs. arXi (2021). https://doi.org/10.48550/ARXIV.2107.12048

9. Kairouz, P., et al.: Advances and open problems in federated learning. arXiv (2019). https://doi.org/10.48550/ARXIV.1912.04977

10. Li, T., Sahu, A., Talwalkar, A., Smith, V.: federated learning: challenges, methods, and future directions. IEEE Signal Process. Mag. **37**(3), 50–60 (2020)

11. scikit-learn. sklearn.cluster.KMeans (2022). https://scikit-learn.org/stable/modules/generated/sklearn.cluster.KMeans.html. Accessed 7 Apr 2022

12. Yang, Z., Chen, M., Saad, W., Hong, C.S., Shikh-Bahaei, M.: Energy efficient federated learning over wireless communication networks. IEEE Trans. Wireless Commun. **20**(3), 1935–1949 (2021). https://doi.org/10.1109/TWC.2020.3037554

13. Pham, Q.-V., Zeng, M., Ruby, R., Huynh-The, T., Hwang, W.-J.: UAV communications for sustainable federated learning. IEEE Trans. Veh. Technol. **70**(4), 3944–3948 (2021). https://doi.org/10.1109/TVT.2021.3065084

14. Cs.toronto.edu. CIFAR-10 and CIFAR-100 datasets (2022). https://www.cs.toronto.edu/kriz/cifar.html

15. He, K., et al.: Deep residual learning for image recognition. arXiv (2015). https://doi.org/10.48550/ARXIV.1512.03385

16. Azari, M.M., Geraci, G., Garcia-Rodriguez, A., Pollin, S.: UAV-to-UAV communications in cellular networks. IEEE Trans. Wireless Commun. **19**(9), 6130–6144 (2020). https://doi.org/10.1109/TWC.2020.3000303

Centralized Multicasting AODV Routing Protocol Optimized for Intermittent Cognitive Radio Ad Hoc Networks

Phetho Phaswana[ID] and Mthulisi Velempini[(⊠)] [ID]

Department of Computer Science, University of Limpopo, Polokwane, South Africa
`mthulisi.velempini@ul.ac.za`

Abstract. The advancement of wireless technology is affected by Spectrum scarcity and the overcrowding of free spectrum. Cognitive Radio Ad Hoc Networks (CRAHNs) have emerged as a possible solution to both the scarcity and overcrowding challenges of the spectrum. The CRAHNs ensure that the Secondary Users (SUs) do co-exist with Primary Users (PUs) in a non-interfering manner. The SUs access the licensed spectrum opportunistically when they are idle. CRAHNs have many use cases which include intermittent networks here referred to as intermittent CRAHNs (ICRAHNs). For example, the Military (MCRAHNs). MCRAHN is complex and characterized by a dynamic topology which is subject to frequent partitioning and route breakages due to attacks and destruction in combat.

This study optimizes the routing protocols for intermittent networks such as the MCRAHNs. ICRAHN routing is a challenge due to the network's intermittent attribute, which is subject to destruction in the case of MCRAHN which is characterized by frequent link breakages. The performance of the proposed routing scheme was evaluated through network simulations using the following metrics: throughput, and Routing Path delay, Node Relay delay, Spectrum Mobility delay. The simulation results show that the MAODV is the best-performing algorithm.

Keywords: Cognitive Radio Ad Hoc Networks · Intermittent Networks · Primary Users · Secondary Users · Spectrum Scarcity

1 Introduction

The emergence of the Fourth Industrial Revolution, the Internet of Things (IoT) and blockchain technologies which require a high-speed network (Internet) connectivity and spectrum have led to spectrum scarcity. Unfortunately, network connectivity depends on the availability of the spectrum and a stable network. To address these challenges of spectrum scarcity, the Federal Communication Commission (FCC) designed a framework which allows secondary users (SUs) to use the licensed spectrum opportunistically when not in use [1].

Routing in intermittent mobile networks is a challenge since there are no guaranteed routing paths. The nodes can be destroyed during the attack while they are relaying packets. This challenge has severe consequences in Intermittent Military Cognitive Radio Ad

J. Taheri et al. (Eds.): MobiCASE 2022, LNICST 495, pp. 31–40, 2023.
https://doi.org/10.1007/978-3-031-31891-7_3

Hoc Networks (IMCRAHNs) nodes such as tankers and aircraft which can be destroyed in combat resulting in the partitioning of the network during a critical phase of the battle. In some cases, routing may be impossible. Longer delays in routing may be incurred resulting in packet-timeout, increased packet drop rate and the degradation of the performance of the network. The delays in IMCRAHNs caused by the destruction of nodes also increase the Routing Path (RP) delay, Spectrum Mobility (SM) delay and Node Relay (NR) delay. The destruction of nodes, therefore, has a ripple effect on the IMCRAHNs. Furthermore, it also affects the achievable throughput as the packet drop rate increases.

The design of routing algorithms in IMCRAHNs requires a dynamic and robust technique which addresses the destruction of nodes and avoids incomplete paths while employing flexible and proactive recovery mechanisms. Several routing protocols exist which are designed to address the IMCRAHNs routing challenges. Unfortunately, current routing algorithms are not optimized for IMCRAHNs routing challenges such as delays. There is a need to optimize routing protocols for Delay Tolerant Networks (DTNs) such as the IMCRAHNs [2]. The routing protocols should reduce delays while improving achievable throughput.

2 Related Work

The routing paths in IMCRAHNs are nondeterministic which degrades the efficiency of routing protocols. Unfortunately, the existing routing protocols are not optimized for IMCRAHNs. We review schemes which were designed to mitigate the effects of SM, RP and NR delays in IMCRAHNs. It was observed that the IMCRAHNs delay is longer than the one for CRAHNs as a result, the IMCRAHNs are categorized as DTNs [3].

The mobility of nodes is also a challenge in ad hoc networks which negatively impacts the performance of routing algorithms. However, the location of nodes, the topology of the ad hoc network and the frequency of changes in the topology determine the routing approach. The design of routing algorithms is also complicated by the size of networks and transmission range. For example, Geo-routing (Geographic routing) is optimized for either geographical or zonal routing. In Geo-routing, packets are broadcasted towards the direction of the zone within which a destination node is likely to be encountered [4].

The Ad Hoc On-Demand Distance vector (AODV) routing algorithm is one of the common MANET routing algorithms [5]. The AODV is being considered for IMCRAHNs and its performance is encouraging. The reactive nature of AODV makes it more suitable for IMCRAHNs which is characterized by dynamic spectrum channel switching.

The Internet Protocol spectrum-aware geographic-based routing protocol (IPSAG) was proposed in [6]. The IPSAG is a geographical and spectrum-aware protocol which employs zonal routing using multicasting. IPSAG relies on prior knowledge of the spectrum and the geographical location of nodes for effective routing. For an effective relay of packets, all the nodes in IPSAG are expected to store the geographical locations of nodes in their neighbourhood or zone. In IPSAG, nodes employ the Greedy forwarding strategy to relay packets according to geographical location information. The nodes forward packets towards the direction of the destination node. The next hop node is expected to be closest to the destination and should have the best spectral quality. If a node has two options to relay packets, spectral density is used as a tie-breaker.

The performance of IPSAG was evaluated against the following routing protocols: The Spectrum Aware Routing for Cognitive Ad-hoc Networks (SEARCH) and the AODV. The results of IPSAG show that it is superior in terms of efficiency. IPSAG incorporates the Common Spectrum Opportunities technique which is used for routing decisions. A node with similar spectrum opportunities to the ones of the relay node is selected for data transmission to avoid channel switching costs [7] and the associated delays.

Though IPSAG was evaluated to be the best protocol, it is likely to drop many packets in intermittent networks with no guaranteed routes. It is not designed to buffer packets until routes are re-established.

The functionality of AODV and its use of sequence numbers to maintain the freshness of routing paths is relevant to IMCRAHNs. It plays a fundamental role in route discovery. When a node receives a Route Request (RREQ), it compares its sequence number to the sequence number of the RREQ. The establishment of the routing path is based on the greater sequence number [8].

Multicasting AODV (MAODV) is a version of AODV and it broadcasts packets to a given segment of the network [9]. However, MAODV does not perform well in repairing routes caused by breakages of relay nodes in IMCRAHNs. In the event of link breakages, the MAODV resumes transmission from the source node instead of continuing from the last relay node.

In WCETT, the best path is selected using the on-demand weighted cumulative expected metric [10]. The routing process is initiated by broadcasting the RREQ. The weighted cumulative transmission time is contained in the RREQ. The RERR is sent when the sequence number of the destination is equal to or less than the one in the route entry. When the RREQ is received, the decision to send an RREP is based on the cost of the RREQ. It should be less than the one of the previous RREQ which has the same sequence number. The paths with the lowest cost are selected.

3 System Model

The algorithms were simulated in three scenarios with 6, 35 and 70 nodes and the simulations were run for 100, 300 and 500 simulation seconds respectively. The simulation times were varied to evaluate effectively the performance of the algorithms. Table 1 presents the values of parameters used in the study [11]. The following metrics were considered: Routing Path delay, Spectrum Mobility delay and Node Relay delay. The metrics are all delay related which are critical in military intermittent networks. Delay tolerate routing schemes are therefore desirable. Therefore, the selection of metrics was informed by a need to reduce delay in IMCRAHNs. Delays in communication in IMCRAHNs may be critical which may result in the loss of life and the destruction of the equipment.

Table 1. The Simulation parameters

Number of Nodes	6, 35, 70
Simulation Time(s)	100 s, 300 s, 500 s
Size of the Packets (bytes)	512
Simulation Grid (m × m)	500 × 500
Traffic Rate/ Rate	Constant Bit Rate (CBR) 4 packets/s
Nodes Velocity (m/s)	12–15
Range of Transmission (m)	90, 120, 150, 180
Number of connections	15, 25, 35
Pause Time (s)	0, 50, 100, 250, 350, 500
Number of Radios	2
Routing Algorithms	AODV, MAODV
Antenna	Omni-directional
MAC Standard	IEEE 802.11b
Number of Pus	6 (For each set of nodes)
Number of SUs	4, 33, 68 (For each set of nodes)

4 Results

We evaluated the effectiveness of the multicasting routing protocol. The MAODV was evaluated and compared to AODV. Figure 1 depicts the RP delay simulation results for MAODV and AODV routing protocols. Figure 1 shows that for scenarios with 6 and 70 nodes, AODV performed poorly in comparison to MAODV. The performance of the AODV is depicted by the maroon curves. The MAODV incurred less RP delay than AODV because it broadcasts packets to a given zone within which the destination node can be reached or a zone closest to the destination node. Within a zone, paths leading to the destination node are selected while broken links are avoided [12].

MAODV is, therefore, a zonal or geographical-based routing protocol, however, in IMCRAHNs, the possibility of route destruction complicates routing. Furthermore, delays and routing overheads are incurred when the whole network is considered for routing. However, the results in Fig. 1 are clustered as a result, we also analyzed the average performance of these schemes in Fig. 2. Furthermore, Fig. 1 is also presented in Appendix A with high resolution.

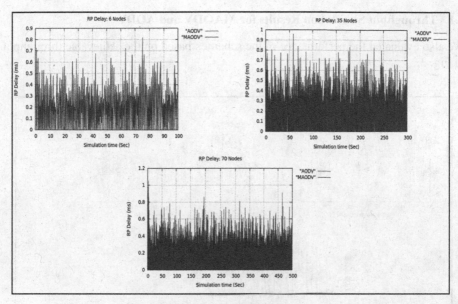

Fig. 1. RP simulation Results

Figure 2 presents the average RP delay results. The average results show that for all the scenarios, the AODV routing protocol experienced more RP delay-related challenges than the MAODV routing protocol. The good performance of the MAODV can be attributed to its effective routing approach discussed under Fig. 1 results and the fact that it is optimized for IMCRAHNs.

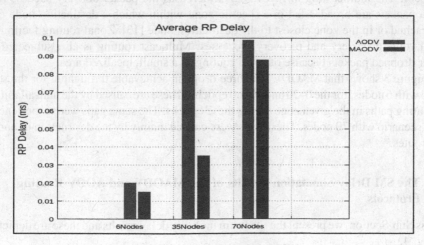

Fig. 2. Average RP delay results

4.1   Throughput Simulation Results for MAODV and AODV

We also evaluated the performance of the schemes based on the achievable throughput in Fig. 3.

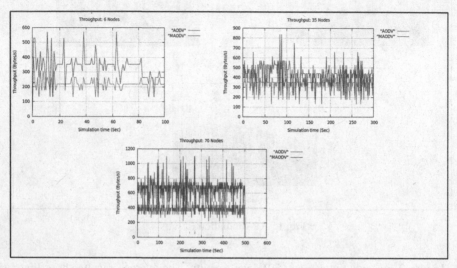

Fig. 3. Throughput Simulation Results

Figure 3 depicts the achievable throughput results of all the network scenarios. The results show that the MAODV achieved more throughput compared to the AODV routing protocol. The multicasting in IMCRAHNs increases the packet delivery success rate when packets are broadcasted in a specific zone within which a destination node can be reached or in the zone closest to the destination node [13]. Zonal routing facilitates faster route discovery and recovery processes. Multicast routing is also subjected to fewer dropped packets because of zonal routing in a small, localized area.

Figure 3 shows that MAODV had three drops in achievable throughput for the scenario with 6 nodes: for the 0–20 and 40–60 epochs. These are caused by the unavailability of routing paths in the given zone during these epochs. The same gaps were experienced for a scenario with 70 nodes. These gaps were caused mainly by the destruction of nodes and routes.

4.2   The SM Delay Simulation Results of the MAODV and AODV Routing Protocols

In this Sub-Section, we present the spectrum mobility delay results and these are depicted in Fig. 4.

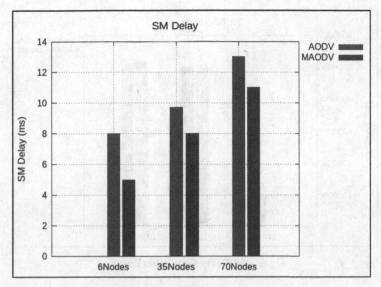

Fig. 4. SM delay simulation Results

Figure 4 shows the SM delay results in which the MAODV was superior to the AODV. Spectrum mobility causes the unavailability of routes and the frequency of this occurrence degrades the performance of the network. However, both MAODV and AODV are impacted negatively by the SM delay. The MAODV is more efficient because it guarantees route availability before the transmission can take place. As a result, the MAODV has a high likelihood of routes being available. Spectrum mobility is a challenge in IMCRAHNs because a channel detected to be available during sensing can become unavailable just before transmission takes place. If this happens, an affected route cannot be used for data transmission. However, this is minimized in IMCRAHNs through the implementation of zonal routing which increases the availability of routing paths for longer periods. As a result, MAODV incurs less SM delay than the AODV routing protocol.

4.3 The NR Delay Simulation Results of the MAODV and AODV Routing Protocols

In this Section, the schemes were evaluated using the Node Relay delay metric and the results are shown in Fig. 5.

Figure 5 shows that the MAODV performs better in all aspects. The MAODV routing protocol incurs the least NR delay compared to the AODV routing protocol because, in IMCRAHNs, zonal routing enables routes to be discovered and repaired faster.

The low NR delay in MAODV is because there is a positive correlation between NR delay and SM delay. For a packet to be relayed, the node first accesses the spectrum. As a result, the factors which affect the SM delay also impact negatively the NR delay. A node therefore, can only relay a packet when the spectrum is available for data transmission [14].

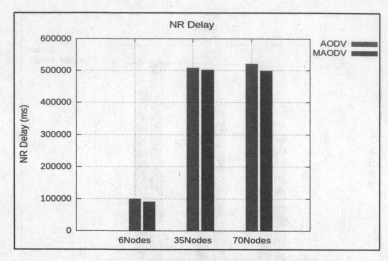

Fig. 5. NR Delay Simulation Results

The results presented in Figs. 1, 2, 3, 4 and 5 show that the MAODV routing protocol is superior to the AODV routing protocol in all the simulation scenarios. The MAODV achieved better results in IMCRAHNs routing largely because of the multicasting technique in a localized and focused zone. In a Multicasting based routing strategy, a network is fragmented logically into smaller zones which contain the destination node or which are closest to the destination node. The relaying of packets is therefore informed by the proximity of the destination node to or within a given zone.

5   Conclusion

The simulation results of the study show that the multicasting routing technique implemented in MCRAHNs is more efficient. Zonal or geographical routing facilitates faster discovery of routing paths while enabling faster recovery of broken routing paths. As a result, the MAODV outperformed AODV.

Figures 1 and 3 also show that for RP delay and throughput simulation results, there were broken routes which were encountered. These are denoted by the drop in achievable throughput in the throughput results. However, despite these challenges of route breakages, the MAODV still performed better. The results show that the MAODV did experience some route breakages which it repaired faster within a given zone.

In the case of SM and NR delay, the results show that the increase in delay is positively correlated with the increase in the number of transmitting nodes. However, in NR and SM delay simulation results, the MAODV routing protocol outperformed the AODV routing protocol. The MAODV routing protocol is more robust and resilient compared to the AODV routing protocol. The implementation of multicasting routing technique ensures that routes in a given zone are available for a longer period which improves MAODV performance. The zonal routing and the use of stable routes reduce SM and NR delays given a higher probability of availability of routing paths for longer durations which in turn, improves the utilization of idle channels.

Acknowledgement. This work was supported by the National Research Foundation of South Africa (Grant Number: 114155).

A Appendix

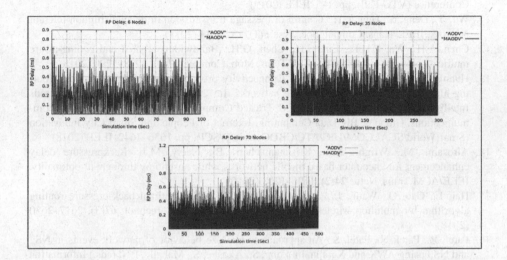

References

1. An, B.: A stability-based spectrum-aware routing protocol in mobile cognitive radio ad-hoc networks. In: 2014 International Symposium on Computer, Consumer and Control, pp. 1014–1017. IEEE (2014)
2. Liu, C., Zhang, G., Guo, W., He, R.: Kalman prediction-based neighbor discovery and its effect on routing protocol in vehicular ad hoc networks. IEEE Trans. Intell. Transp. Syst. **21**(1), 159–169 (2019)
3. Tang, F., et al.: Delay-minimized routing in mobile cognitive networks for time-critical applications. IEEE Trans. Industr. Inf. **13**(3), 1398–1409 (2017)
4. Huang, J., Wang, S., Cheng, X., Liu, M., Li, Z., Chen, B.: Mobility-assisted routing in intermittently connected mobile cognitive radio networks. IEEE Trans. Parallel Distrib. Syst. **25**(11), 2956–2968 (2014)
5. Zhang, L., Zhuo, F., Bai, C., Xu, H.: Analytical model for predictable contact in intermittently connected cognitive radio ad hoc networks. Int. J. Distrib. Sens. Netw. **12**(7), 1–12 (2016)
6. Ghafoor, H., Koo, I.: Spectrum and connectivity aware anchor-based routing in cognitive vehicular ad hoc networks. In: 2016 Eighth International Conference on Ubiquitous and Future Networks (ICUFN), pp. 679–684. IEEE (2016)
7. Velempini, M., Dlodlo, M.E.: A multiple channel selection and coordination MAC scheme. In: 2009 Second International Conference on Advances in Mesh Networks, pp. 126–131 (2009). https://doi.org/10.1109/MESH.2009.30

8. Landmark, L., Larsen, E., Hauge, M., Kure, Ø.: Resilient internetwork routing over heterogeneous mobile military networks. In: MILCOM 2015–2015 IEEE Military Communications Conference, pp. 388–394. IEEE (2015)
9. Vang, Y., Saavedra, A., Yang, S.: Ferry enhanced improved PRoPHET routing protocol. In: 2015 IEEE 12th International Conference on Mobile Ad Hoc and Sensor Systems, pp. 567–572. IEEE (2015)
10. Melvin, S., Lin, J., Kim, S., Gerla, M.: Hitchhiker: a wireless routing protocol in a delay tolerant network using density-based clustering. In: 2018 IEEE 88th Vehicular Technology Conference (VTC-Fall), pp. 1–5. IEEE (2019)
11. Wu, Y., Deng, S., Huang, H.: Control of message transmission in delay/disruption tolerant network. IEEE Trans. Comput. Soc. Syst. **5**(1), 132–143 (2018)
12. Chen, Y.H., Wu, E.H.K., Lin, C.H., Chen, G.H.: Bandwidth-satisfied and coding-aware multicast protocol in MANETs. IEEE Trans. Mob. Comput. **17**(8), 1778–1790 (2018)
13. Huang, W., Ma, Z., Dai, X., Xu, M.: Connectivity probability based spray and wait routing algorithm in mobile opportunistic networks. In: 2018 IEEE SmartWorld, Ubiquitous Intelligence & Computing, Advanced & Trusted Computing, Scalable Computing & Communications, Cloud & Big Data Computing, Internet of People and Smart City Innovation (SmartWorld/SCALCOM/UIC/ATC/CBDCom/IOP/SCI), pp. 1636–1642. IEEE (2018)
14. Alresaini, M., Wright, K.L., Krishnamachari, B., Neely, M.J.: Backpressure delay enhancement for encounter-based mobile networks while sustaining throughput optimality. IEEE/ACM Trans. Netw. **24**(2), 1196–1208 (2016)
15. Hai, L., Gao, Q., Wang, J., Zhuang, H., Wang, P.: Delay-optimal back-pressure routing algorithm for multihop wireless networks. IEEE Trans. Veh. Technol. **67**(3), 2617–2630 (2018)
16. Patel, R., Patel, N., Patel, S.: An approach to analyze behavior of network events in NS2 and NS3 using AWK and Xgraph. In: Fong, S., Akashe, S., Mahalle, P.N. (eds.) Information and Communication Technology for Competitive Strategies. LNNS, vol. 40, pp. 137–147. Springer, Singapore (2019). https://doi.org/10.1007/978-981-13-0586-3_14
17. Prashanthi, G., Rao, A.V.: A survey on relay selection for geographical forwarding in sleep-wake cycling wireless sensor networks. Int. J. Res. **5**(7), 229–237 (2018)
18. Malhotra, S., Trivedi, M.C.: Authentication, KDC, and key pre-distribution techniques-based model for securing AODV routing protocol in MANET. In: Panigrahi, B., Trivedi, M., Mishra, K., Tiwari, S., Singh, P. (eds.) Smart Innovations in Communication and Computational Sciences. Advances in Intelligent Systems and Computing, vol. 669, pp. 175–186. Springer, Singapore (2019). https://doi.org/10.1007/978-981-10-8968-8_15
19. Jhajj, H., Datla, R., Wang, N.: Design and implementation of an efficient multipath AODV routing algorithm for MANETs. In: 2019 IEEE 9th Annual Computing and Communication Workshop and Conference (CCWC), Las Vegas, NV, USA, pp. 0527–0531 (2019)

Machine Learning/Deep Learning

YouBrush: Leveraging Edge-Based Machine Learning in Oral Care

Esteban Echeverri[1]([envelope]) [iD], Griffin Going[1] [iD], Rahat Ibn Rafiq[1] [iD],
Jonathan Engelsma[1] [iD], and Venu Vasudevan[2] [iD]

[1] Grand Valley State University, Allendale, MI, USA
{echevere.mail,goinggr,rafiqr,engeljo}@gvsu.edu
[2] P&G,Cincinnati, USA
vasudevan.v@pg.com

Abstract. A disconnect is frequent regarding the length of time a person claims to have brushed their teeth and the actual duration; the recommended brushing duration is 2 min. This paper seeks to bridge this particular disconnect. We introduce YouBrush,—a low-latency, low-friction, and responsive mobile application—to improve oral care regimens in users. YouBrush is an IOS mobile application that democratizes features previously available only to intelligent toothbrush users by incorporating a highly accurate deep learning brushing detection model—developed by Apple's createML—on the device. The machine learning model, running on the edge, allows for a low-latency, highly responsive scripted-coaching brushing experience for the user. Moreover, we craft in-app gamification techniques to further user interaction, stickiness, and oral care adherence.

Keywords: Mobile Development · Oral Care · Sound

1 Introduction

Mobile phones and embedded sensors are becoming increasingly influential in various health care use-cases. Several researchers have demonstrated the success of using sensor technologies to monitor the exercise, dietary, and sleep regimen of subjects [7,38]. This paper uses mobile phones' audio-sensory capability to improve oral care regimens. Oral health care is of paramount importance—researchers report it significantly impacts one's quality of life [15]. Notwithstanding the proven impact of oral care on physical and emotional well-being [15], studies demonstrate that a substantial portion of the population brushes with an incorrect technique, such as improper toothbrushing [13].

Several mobile applications and researches focusing on healthcare have come to the fore [12] in recent years. The majority of these researches leverage optical motion capture systems [18], accelerometer sensors installed in smart toothbrushes [20], and audio sensors [21]. Positive feedback and gamification techniques such as leaderboards have been influential in improving subjects' daily oral care

J. Taheri et al. (Eds.): MobiCASE 2022, LNICST 495, pp. 43–58, 2023.
https://doi.org/10.1007/978-3-031-31891-7_4

regimen of subjects [12, 22]. Other oral care use-cases such as disease detection [5, 36] have seen a significant application of deep learning approaches. Researches have shown that real-time positive feedback and a highly responsive coaching mechanism can significantly improve a user's experience with a mobile application that seeks to improve a particular behavior [10, 33]. To this aim, in this paper, we present YouBrush, which employs edge-based deep learning, mobile phones' audio sensory capability, and a gamification mechanism to facilitate the toothbrushing regimen of a subject. First, we collect and label brushing audio data, and bathroom ambient sounds to construct a robust and diverse training dataset. Then, we use Apple's off-the-shelf createML [1] to train and develop a deep learning model that accurately detects brushing events using the dataset. Next, we implement a gamification and dynamic feedback mechanism to augment user experience and engagement. Finally, we incorporate the deep learning and gamification modules in the YouBrush mobile application, which we develop in Swift programming language [4]. The edge-based brushing audio detection enables YouBrush to provide a seamless, highly responsive scripted-coaching brushing experience for the user—facilitating an engaging user experience.

Concerning other oral care aid applications with specific deference to non-smart toothbrushes, a weakness is present in the user trust requirement. Such applications often, for example, present themselves as a simple two-minute timer when in the context of an actual brushing/oral care session. Thus, the application has little power to ensure that the user spends the entire two-minute period brushing their teeth. Instead, it must simply assume that the user does what they say they are doing or will do. The aforementioned disconnect between user perception and reality makes this a less-than-ideal implementation. Of the primary goals of YouBrush, one of the most important is to replace this requirement with an accountability mechanic. This is done by using the edge-based brushing audio detection to drive a two-minute timer, forcing the user to actively participate in the brushing activity to push the timer forward instead of simply observing the timer. YouBrush relies heavily on this inversion, creating an environment in which the user's actions and habits drive application progress and logic where viable instead of the application solely driving the user.

Attempts such as YouBrush that log and evaluate daily activity are critical in promoting a healthy lifestyle [21]. Fitbit[1], an activity-tracker company in the wearable technology space, journals a person's running, walking, sleeping habit, and heart rate, for example. This information-logging is a tremendous help for the user to self-evaluate their daily progress. We argue that YouBrush has the potential to serve a similar requirement to evaluate toothbrushing performance. Moreover, the presence of a competitive leaderboard can motivate the users further to be stricter followers of daily oral-care regimens.

We organize the rest of the paper as follows. *First*, we introduce the related work in four subareas related to our research: oral care and mobile applications, oral care and machine learning, oral care and gamification, and brushing sensing. *Second*, we present the YouBrush application design and implementation details.

---

[1] https://www.fitbit.com/global/us/home.

Third, we describe the methodologies used to collect, augment and label the audio datasets we use to train our deep learning model to detect brushing events. *Finally*, we detail the tools and techniques used to train and implement the deep learning model and present the performance results.

2 Related Work

2.1 Oral Health Care and Mobile Apps

Mobile-based healthcare applications have spiked in recent years, and oral healthcare is no exception. Dentists, orthodontists, and oral health care evangelists have sought to improve their patients' oral care regimens by leveraging this ubiquitous nature of the mobile paradigm. Such efforts include health-risk information provision, self-monitoring of behavior and behavioral outcomes, prompting barrier identification, setting action and coping plans, and reviewing behavioral goals [30]. Researchers have also proposed mobile-based solutions to focus on diseases and treatments, such as oral mucositis [23], facilitation of the recovery process of patients who just went through an orthodontic treatment [37], promotion of oral hygiene among adolescents going through such treatments [34]. Recent works have also put under the microscope ways to ensure regular engagement of oral-care-related mobile apps—for example, motivating users to brush their teeth for 2 min using music [40].

2.2 Oral Care and Gamification

Gamification is defined as applying game design elements in non-game contexts [11]. Several Oral Care mobile applications employ gamification techniques such as badges, leaderboards, and levels to increase user engagement and activity [12]. In addition, researchers have recently investigated the effect of positive feedback through mobile applications and gamification techniques to gauge the improvement of oral care regimens among subjects [22].

2.3 Brushing Sensing

A recent study has found that real-time feedback can significantly improve the quality of brushing [18]. In [18], the authors evaluated the ability of a power brush with a wireless remote display to improve brushing force and thoroughness. Optical motion capture systems [18], accelerometer sensors embedded in brushing devices [20,39], and brushing audio sounds [21] have also been leveraged to analyze brushing behaviors. Research has shown that proper graphical user feedback during brushing—a cartoon display for children to show regions of the mouth that were adequately brushed to be displayed as plaque-free [6] or smart digital visual toothbrush monitoring and training system (DTS) in terms of correct brushing motion and grip axis orientation [16]—can motivate better oral care regimen. In [21], the authors proposed a low-cost system built around an off-the-shelf smartphone's microphone to evaluate toothbrushing performance using hidden Markov models. However, in this paper, we leverage gamification and deep learning in addition to audio sensors to evaluate brushing performance. Audio and deep learning, in recent years, have seen several applications in areas such as speech [8], music [25], and environmental sound detection [27].

3   Application Design and Implementation

YouBrush revolves around three primary aims: Ensure regular brushing, quality of brushing technique, and every brushing event lasts at least 2 min [31]. For this purpose, we proceed to the application design and implementation phase of YouBrush. We use Swift to develop the frontend and use Firebase[2] as the backend of our application. Next, we focus on four design choices imperative for the implementation phase of YouBrush: handling live audio, in-app data presentation, privacy and latency concerns, and constructing the in-app game mechanics. The following subsections describe the implementation of each of the above-mentioned design choices (Table 1).

3.1   Audio Processing

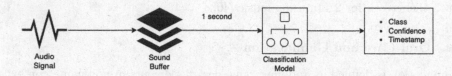

Fig. 1. Audio pipeline

Figure 1 depicts the pipeline we develop to process the audio stream whenever the microphone detects an audio signal. First, a sound buffer stores the audio stream and chunks them into 1-s audio—this chunking helps us later in the audio classification phase. Second, the pipeline sends the 1-s audio samples to the classification module for brushing event detection, enabling the app to accurately timestamp a brushing event's start and end point.

3.2   GUI Design

Brushing Screen: Figure 1a depicts the brushing screen. This screen serves the following. First, it provides the user with immediate goals after a brushing event is initiated. The immediate goals represent different stages of a brushing session—suggesting to the user which areas of their teeth (top and bottom of their teeth, for example, in Fig. 1a) they should be brushing. These immediate

---

[2] https://firebase.google.com/.

Table 1. YouBrush Application

| (a) Brushing Screen | (b) Dashboard | (c) Leaderboard |

goals represent real-time scripted coaching to improve brushing quality. Second, when the user finally completes the final goal, the app marks the brushing session as complete, ensuring a two-minute brushing session.

Dashboard Screen: Figure 1b shows the dashboard—the screen the users see after they log in. We highlight four points on the screen. The brown and pink-colored teeth at the bottom denote an incomplete (or did not use the app when brushing) and complete brushing session at the specified time, respectively. On the top, a timer lets the user know when is their next brushing session. The pink-colored tooth represents how clean their teeth are at the moment—an estimation based on their last completed brushing event. Finally, at the very bottom, the app provides a total score that we calculate based on user consistency. We use this score to facilitate the in-app game mechanics—to be described in Sect. 3.4.

3.3 Privacy and Latency Concerns

We implement the audio processing pipeline on edge to ensure minimal latency and privacy concerns. This design choice allows us an efficient minimal-latency real-time scripted coaching mechanism when the user initiates a brushing event,—an attribute imperative for an engaging user experience—and also enables us to bypass the users' privacy concerns. We do not keep audio from the brushing sessions on the server; all the audio analysis is performed on the device. Additionally, no audio is saved or stored on the user's device beyond the circular eight-kilobyte buffer from which the model draws audio to classify, which itself is not preserved beyond the scope of the view in which it resides. Table 2 lists all the information we collect for a particular user. The set of data that we collect can be divided into two types: information related to a brushing event, such as its timestamp, and information related to the in-app game mechanics.

Table 2. Backend Values

Value	Usage
firstBrushingTime	time of the first (often morning) brushing session. Default: 8AM
secondBrushingTime	time of the second (often midday) brushing session. Default :2PM
thirdBrushingTime	time of the third (often night) brushing session. Default :10PM
brushingStreak	consecutive number of completed brushing sessions
dailyStreak	consecutive number of days for which all brushing sessions were completed.
division	Leaderboard division the user is currently in
totalScore	total score accumulated by the user (lifetime)
weeklyScore	weekly score accumulated by the user

3.4   Gamification

Gamification techniques such as leaderboards and levels have been shown to increase user engagement in oral care apps in several recent studies [12]. For this purpose, we implement a leaderboard functionality in YouBrush, as depicted in Fig. 1c. We give the user a score based on the duration and regularity of their brushing – both daily and across days – and rank them in divisions against simulated users [19]. The leaderboard view in Fig. 1c shows a user in their current position within a leaderboard full of simulated users; these simulated users scale upwards in competitive difficulty as they ascend to higher divisions. Note that the leaderboard view is personalized; every user's view of the leaderboard and the set of simulated users differs from everyone else.

4   Data Collection and Labeling

The organization of this section is as follows. First, we delineate the methodology applied to create the brushing audio dataset used to train and evaluate the eventual machine learning algorithm for brushing detection. Second, we discuss the ethical considerations of collecting audio data for our research. Finally, we introduce a labeling tool we develop to facilitate audio labeling using video recordings.

4.1   Methodology

In addition to recording the team members' audio while brushing—a dataset that we will refer to as self-owned data from now on—we also take advantage of two public sources of audio data: Audioset [14] and Freesound.org [3]. Google's audioset is a large-scale collection of human-labeled 10-s sound clips drawn from YouTube videos. To collect all the audio data, the authors worked with human annotators who verified the presence of sounds they heard within YouTube segments. To nominate segments for annotation, they relied on YouTube metadata and content-based search [14]. Freesound is a collaborative repository of CC-licensed audio samples and a non-profit organization with more than 500,000 sounds and 8 million registered users. Sounds are uploaded to the website by its users and cover a wide range of subjects, from field recordings to synthesized sounds [3]. We ensure to collect audio samples labeled as different classes of bathroom-ambient sounds such as sink, tub, toilet, shower, negative, and silence. We define the tag *silence* as little or no sound and *negative* as any sound samples that do not belong to the labels mentioned above. We provide a detailed description of the audio classes in Sect. 5. Table 3 presents detailed statistics of our dataset, specifying individual audio classes and the length of audio samples for them.

Table 3. Dataset statistics

Class	Number of files	Total Audio Samples	Audio length	Source
Brushing	205	3945	65.75 min	Self-owned and freesound.org
Silence	49	488	8.133 min	Audioset
Sink	68	1478	24.633 min	Self-owned
Toilet	69	387	6.45 min	Self-owned
Tub	20	1522	25.367 min	Self-owned
Negative	25	2046	34 1 min	Self owned
Shower	14	1811	30.183 min	Self-owned

4.2 Ethical Consideration

Crawling and analyzing data from a private environment such as the bathroom entails serious ethical implications. However, in this research, we only concern ourselves with publicly available audio data sets and refrain from interacting with human subjects besides the ones on our team. Therefore, this research conforms to the standard ethics guidelines to protect users [32].

4.3 Labeling

We implement an audio-video labeling tool to construct the self-owned dataset, as shown in Fig. 2. A typical usage flow of the labeling tool is as follows. First, the subject uses their mobile phone to videotape their brushing event. The subject then uploads the video recording to the labeling tool. Upon uploading, the user can move the scrubbing slider to select a portion of the video corresponding to a class of audio specified in the drop-down list located just below the slider. Next, the user clicks "export segments", which prompts the tool to extract just the audio segment from the video file along with the associated label, duration of the audio segment, and start and end timestamp in the parent video file. The purpose of this labeling tool is threefold. First, The labeling tool allows us to bypass any need to store videos of users brushing on the server; we did not store any videos of the users participating in the data collection process. Instead, we only collect the audio extracted from the brushing videos, thereby protecting the users' privacy accordingly [42]. Second, The tool allows us to label the brushing audios with added granularity for future brushing behavior analysis. For example, one particular brushing sub-event audio can be tagged by the user as brushing the front teeth or the back teeth, using the "data label" drop-down list shown in Fig. 2. Third, the tool allows the audio labels to be as accurate as possible. Brushing is a continuous event that ranges from 40 s to two minutes or more, containing not just brushing sounds but sometimes silences, bathroom ambient sounds, or even human speech. The tool's scrubbing slider allows us to extract the exact audio segment containing a brushing event, devoid of any other foreign audios, as much as possible.

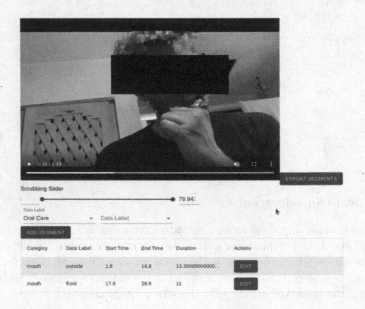

Fig. 2. Labeling Tool used to construct self-owned brushing dataset

5 Machine Learning Implementation

As mentioned in Sect. 3, YouBrush listens to the live audio when the user initiates a brushing event. We present a two-phase classification engine to ensure efficient processing and effective classification of the live audio stream. In the first phase, we constantly take a fixed size of 1 s from the live audio chunk-stream and use our Machine Learning Classifier (MLC) to classify them independently.

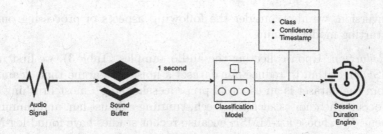

Fig. 3. Second Phase

The second phase uses the classification labels—instead of the audio—to infer the total duration of the user's brushing event. Figure 3 shows our two-phase classification engine. Domestic environments produce a wide variety of audio sounds [35] and a brushing event is no exception. This challenge necessitates us to consider—to train and construct our audio MLC—other surrounding sounds such as running-shower, running-sink, etc. In addition to brushing audio. Table 4 lists the classes of audio we consider to develop our brushing MLC.

Table 4. Classes of Audio Considered

Label	Description
Brushing	Traditional Toothbrush Brushing
Shower	Shower Running
Sink	Sink Running
Toilet	Toilet Flushing
Tub	Tub Filling
Silence	Plain Silence
Negative	Others (Fan Running etc.)

We organize the rest of the section into two subsections. The first sub-section, *Preprocessing*, delineates the approaches we follow to pre-process the audio samples to construct the training data. The second sub-section, *Model Creation*, describes and evaluates our brushing MLC.

5.1 Preprocessing

We convert all sound files to single channel wav files with a sampling rate of 16 kHz. We provide two reasons for this conversion choice. First, researchers have found such conversions effective in environmental sound classification tasks using Deep Learning Models [9]. Second, to develop our brushing MLC, we will use Apple's off-the-shelf CreateML platform (described in more detail in Sect. 5.2); and createML recommends audio file's sample rate to be 16 kHz [2]. Besides the rate conversion, we also consider the following aspects of processing our data before starting model training.

Noise Reduction: Upon collecting the audio samples (Table 3), we first use the log-MMSE algorithm to reduce our dataset's noise. We argue that cleaning the noise from the dataset is an essential preprocessing step—most brushing events tend to occur when noises such as speech, running exhaust fan, and running sink are present. We choose log-MMSE because recent studies have found log-MMSE to be significantly effective in reducing noise when the actual noise is unknown [17,41].

Data Augmentation: One of the main drawbacks of deep learning approaches is the need for a significant amount of training data [24]. Therefore, researchers have turned to data augmentation as the go-to approach to tackle the training data to implement deep learning models [28,29]. We leverage Python's librosa library [26] to implement the data augmentation step when creating the training data. Table 5 lists the augmentation techniques we leverage to construct our training dataset.

Table 5. Audio Augmentation Techniques Performed

Method	Description
Pitch Modification	Using a factor to modify pitch
Stretching	Time-stretch an audio series by a fixed rate.
Changing Decibels	Modifying the decibels from an audio file, changing how loud it sounds
Noise Injection	Combining two different audios from the original dataset to get a new sample, for example combining sink running audio files with brushing files

MFCC Features: To visually analyze audio samples and engineer features for model training, we choose Mel-Frequency Cepstral Coefficient (MFCC) [43]. Mel-Frequency Cepstrum (MFC) is a representation of the short-term power spectrum of a sound based on a linear cosine transform of a log power spectrum on a nonlinear Mel scale of frequency. MFCCs are coefficients that collectively make up an MFC. They are derived from a type of cepstral representation of the audio clip (a nonlinear "spectrum-of-a-spectrum") [43]. MFCC as audio features have outperformed Fourier Transform (FT), Homomorphic Cepstral Coefficients (HCC), and others for non-speech audio use-cases [43].

5.2   Model Creation

Upon completing the audio preprocessing and preliminary analysis, we partition our dataset into training and test data, using an 80–20 split. Additionally, we

Table 6. Model results

Class	Precision %	Recall %
Brushing	97	92
Shower	100	96
Sink	95	95
Toilet	83	71
Tub	90	100
Silence	77	76
Negative	75	97

inject a large amount of publicly-sourced audio data into the test set to ensure testing for generalization at a high level. We use Apple's off-the-shelf tool CreateML to create and train custom machine learning models [1]. CreateML can be used for various use cases, such as recognizing images, classifying audios, extracting meaning from text, or finding relationships between numerical values. It allows for modifying a few aspects of the metadata during model creation (Maximum number of iterations and the overlap factor) and creates a coreml file to be used on any swift application. Because we use Swift to develop YouBrush (Sect. 3), Table 6 shows the precision and recall results on our test data. We use the CreateML-setup to take at most 25 iterations to finish the training and use an overlap factor of 50%. One characteristic of CreateML is that it configures all the internal setup of the model, for example, how many layers it will have and what type of layer it will use; everything is based on the use-case selected. We select the use-case of Sound Classification. After the training phase, the deep learning model has the following configuration: 6 convolution layers, 6 ReLU activation layers, 4 Pooling layers, and one flattened layer, followed by the softmax layer. Table 6 shows the precision and recall of the model's performance for every audio class in the test dataset. As it can be seen, for the brushing class, the model achieves precision and recall scores of 97% and 97%, respectively, using MFCC features.

5.3 Model Testing

Recognizing that the physical domain in which the model operates is home to many potential confounding sounds, we give special consideration to injecting such sounds into the training data as background noise to ensure we account for these. Additionally, the "Negative" class is composed of such sounds (e.g., bathroom exhaust fans, speech) to better define the boundary betwixt the brushing class and such confounding sounds concerning their tendency to occur in tandem. Finally, when injecting publicly-sourced data into our test sets, we give special deference to those where common confounding sounds are dominant.

Of particular concern is the evaluation scenarios by which we must test the model. Table 6 displays testing results concerning individual samples, wherein a sample is one single second of audio (achieved by decomposing full audio samples of greater length). These results offer a valuable evaluation of the model at a glance when quickly iterating, but of greater importance is the distribution of these results as they relate to samples of greater length. Consider the following: we sample ten seconds of a user engaged in their brushing session. Let's assume we classify seven seconds as brushing and wrongly classify the rest as some other arbitrary class. Based on the YouBrush implementation, the way these three misclassified seconds are distributed throughout the sample is significant. Namely, suppose these misclassified second-long samples occur disjointly from one another (e.g., are separated by correctly classified periods). In that case, the interruption to the user experience is lessened and can even be smoothed over by assuming certain levels of uncertainty (which we will discuss shortly). If the inverse occurs and these misclassified second-long samples occur consecutively with respect to time (e.g., are not separated by any period of properly-classified brushing samples), the user experience is degraded, and both trust and interest in the experience provided by the application quickly break down. For this reason, we separate model testing into two focuses: traditional sample-based testing, in which we focus on achieving accuracy for singular samples, and longer-form implementation-based testing, in which we focus on accuracy over time for sets of ordered samples. Within implementation-based tests, more extended periods of inaccuracy are more heavily penalized. Implementation-based tests are built to model real-life brushing sessions and therefore focus on improving the user experience.

5.4   Model Implementation

Finally, we address the subject of model uncertainty and how YouBrush addresses such shortcomings at the implementation level. Recognizing that failing closed (e.g. failing to properly classify moments in which the user is brushing) creates a negative experience for the user, the YouBrush implementation has a set of built-in fail-safes to make failing opening (e.g. considering moments of uncertainty to be aurally contiguous with a set of previously classified samples) the default in uncertain periods of sound. The first of these is simple: the brushing class does not require a high level of confidence. This allows samples nearer the class boundary to nonetheless be considered as truth.

The second of these is a code-level concession offered to the model based on the assumption that each short-term sample taken (ranging from periods of a half-second to a full second) is aurally similar to those before and thus does not represent a significant departure from the pattern established in a small set of previous samples on their own (with respect to the order in which samples are taken). Based on this assumption, we define a variable amount of time that a departure from a previously established pattern must meet to break the said pattern. Within the YouBrush-specific implementation, only the brushing sound class is of interest, and thus the only pattern eligible is that of consecutive brushing sounds. Therefore, once a short pattern of brushing sound is established, model uncertainty (by way of misclassification of brushing audio) must occur consecutively several times before the user experience is adversely affected. Returning to the previous example wherein seven of ten seconds of audio are properly classified as brushing, with the other three incorrectly classified as a non-brushing class, we address both hypothetical distributions of these incorrectly classified samples wherein they occur either disjointedly or as a set; in both circumstances, the threshold for pattern departure is not met, and model uncertainty is abstracted away from the user experience.

6 Conclusion

In this paper, we denote our experience on how to develop a mobile application, YouBrush. We delineate the design choices we make to ensure a real-time scripted coaching brushing experience for the users with traditional toothbrushes, using an on-device highly accurate, low-latency deep learning model for brushing audio detection. To ensure the machine learning model's robustness, we utilize an in-house labeling tool that facilitates precise audio labeling through video-ques. Next, to encourage consistency, regularity, and user engagement, we incorporate in-app game mechanics and real-time positive feedback mechanisms during brushing. In the future, we plan to release YouBrush in the AppStore and investigate its efficacy among the users, performing elaborate user behavior analysis.

Acknowledgements. This work would not be possible without the support of Daniel Londoño, who helped us with our UI design, and Evan Boardway and Meghan Harris, who helped to implement the logic and design for the Leaderboard.

References

1. createml. https://developer.apple.com/documentation/createml. Accessed Apr 2022
2. createml sound classifier. https://developer.apple.com/documentation/createml/mlsoundclassifier. Accessed Apr 2022
3. freesound.org. https://en.wikipedia.org/wiki/Freesound. Accessed 25 July 2022
4. Swift programminmg language. https://developer.apple.com/swift/. Accessed 28 July 2022
5. Camalan, S., et al.: Convolutional neural network-based clinical predictors of oral dysplasia: class activation map analysis of deep learning results. Cancers **13**(6), 1291 (2021)
6. Chang, Y.C., et al.: Playful toothbrush: ubicomp technology for teaching tooth brushing to kindergarten children. In: Proceedings of the SIGCHI Conference on Human Factors in Computing Systems, pp. 363–372 (2008)
7. Chen, Z., et al.: Unobtrusive sleep monitoring using smartphones. In: 2013 7th International Conference on Pervasive Computing Technologies for Healthcare and Workshops, pp. 145–152. IEEE (2013)

8. Chiu, C.C., et al.: State-of-the-art speech recognition with sequence-to-sequence models. In: 2018 IEEE International Conference on Acoustics, Speech and Signal Processing (ICASSP), pp. 4774–4778. IEEE (2018)

9. Chong, D., Zou, Y., Wang, W.: Multi-channel convolutional neural networks with multi-level feature fusion for environmental sound classification. In: Kompatsiaris, I., Huet, B., Mezaris, V., Gurrin, C., Cheng, W.-H., Vrochidis, S. (eds.) MMM 2019. LNCS, vol. 11296, pp. 157–168. Springer, Cham (2019). https://doi.org/10.1007/978-3-030-05716-9_13

10. Croyère, N., Belloir, M.N., Chantler, L., McEwan, L.: Oral care in nursing practice: a pragmatic representation. Int. J. Palliat. Nurs. 18(9), 435–440 (2012)

11. Deterding, S., Dixon, D., Khaled, R., Nacke, L.: From game design elements to gamefulness: defining "gamification". In: Proceedings of the 15th International Academic MindTrek Conference: Envisioning Future Media Environments, pp. 9–15 (2011)

12. Fijačko, N., et al.: The effects of gamification and oral self-care on oral hygiene in children: systematic search in app stores and evaluation of apps. JMIR Mhealth Uhealth 8(7), e16365 (2020)

13. Ganss, C., Schlueter, N., Preiss, S., Klimek, J.: Tooth brushing habits in unstructured adults-frequency, technique, duration and force. Clin. Oral Invest. 13(2), 203–208 (2009)

14. Gemmeke, J.F., et al.: Audio set: an ontology and human-labeled dataset for audio events. In: Proceedings of IEEE ICASSP 2017. New Orleans, LA (2017)

15. Gerritsen, A.E., Allen, P.F., Witter, D.J., Bronkhorst, E.M., Creugers, N.H.: Tooth loss and oral health-related quality of life: a systematic review and meta-analysis. Health Qual. Life Outcomes 8(1), 1–11 (2010)

16. Graetz, C., et al.: Toothbrushing education via a smart software visualization system. J. Periodontol. 84(2), 186–195 (2013)

17. Hu, Y., Loizou, P.C.: Subjective comparison and evaluation of speech enhancement algorithms. Speech Commun. 49(7), 588–601 (2007). https://doi.org/10.1016/j.specom.2006.12.006, https://www.sciencedirect.com/science/article/pii/S0167639306001920. speech Enhancement

18. Janusz, K., Nelson, B., Bartizek, R.D., Walters, P.A., Biesbrock, A.R.: Impact of a novel power toothbrush with smartguide technology on brushing pressure and thoroughness. J. Contemp. Dent. Pract. 9(7), 1–8 (2008)

19. Karatassis, I., Fuhr, N.: Gamification for websail. In: GamifIR@ SIGIR (2016)

20. Kim, K.S., Yoon, T.H., Lee, J.W., Kim, D.J.: Interactive toothbrushing education by a smart toothbrush system via 3D visualization. Comput. Methods Programs Biomed. 96(2), 125–132 (2009)

21. Korpela, J., Miyaji, R., Maekawa, T., Nozaki, K., Tamagawa, H.: Evaluating tooth brushing performance with smartphone sound data. In: Proceedings of the 2015 ACM International Joint Conference on Pervasive and Ubiquitous Computing, pp. 109–120 (2015)

22. Kumarajeewa, R., Jayarathne, P.: Improving children's oral hygiene habits in Sri Lanka via gamification. Asian J. Manag. Stud. 2(1), 98–113 (2022)

23. Lin, T.H., Wang, Y.M., Huang, C.Y.: Effects of a mobile oral care app on oral mucositis, pain, nutritional status, and quality of life in patients with head and neck cancer: a quasi-experimental study. Int. J. Nurs. Pract. 28(4), e13042 (2022). https://doi.org/10.1111/ijn.13042

24. Marcus, G.: Deep learning: a critical appraisal. arXiv preprint arXiv:1801.00631 (2018)

25. McFee, B., Ellis, D.P.: Better beat tracking through robust onset aggregation. In: 2014 IEEE International Conference on Acoustics, Speech and Signal Processing (ICASSP), pp. 2154–2158. IEEE (2014)

26. McFee, B., et al.: librosa: audio and music signal analysis in python. In: Proceedings of the 14th python in science conference, vol. 8, pp. 18–25. Citeseer (2015)

27. Mesaros, A.: Detection and classification of acoustic scenes and events: outcome of the DCASE 2016 challenge. IEEE/ACM Trans. Audio Speech Lang. Process. **26**(2), 379–393 (2017)

28. Mushtaq, Z., Su, S.F.: Environmental sound classification using a regularized deep convolutional neural network with data augmentation. Appl. Acoust. **167**, 107389 (2020)

29. Nanni, L., Maguolo, G., Paci, M.: Data augmentation approaches for improving animal audio classification. Eco. Inform. **57**, 101084 (2020)

30. Patil, S., et al.: Effectiveness of mobile phone applications in improving oral hygiene care and outcomes in orthodontic patients. J. Oral Biol. Craniofac. Res. **11**(1), 26–32 (2021). https://doi.org/10.1016/j.jobcr.2020.11.004, https://www.sciencedirect.com/science/article/pii/S2212426820301652

31. Raypole, C.: 5 toothbrushing FAQs. https://www.healthline.com/health/how-long-should-you-brush-your-teeth Accessed 13 July 2022

32. Rivers, C.M., Lewis, B.L.: Ethical research standards in a world of big data. F1000Research, **3**(38), 38 (2014)

33. Schäfer, F., Nicholson, J., Gerritsen, N., Wright, R., Gillam, D., Hall, C.: The effect of oral care feed-back devices on plaque removal and attitudes towards oral care. Int. Dent. J. **53**(S6P1), 404–408 (2003)

34. Scheerman, J.F.M., et al..: The effect of using a mobile application ("WhiteTeeth") on improving oral hygiene: a randomized controlled trial. Int. J. Dent. Hyg. **18**(1), 73–83 (2020). https://doi.org/10.1111/idh.12415. https://onlinelibrary.wiley.com/doi/abs/10.1111/idh.12415

35. Serizel, R., Turpault, N., Eghbal-Zadeh, H., Shah, A.P.: Large-scale weakly labeled semi-supervised sound event detection in domestic environments. arXiv preprint arXiv:1807.10501 (2018)

36. Sultan, A.S., Elgharib, M.A., Tavares, T., Jessri, M., Basile, J.R.: The use of artificial intelligence, machine learning and deep learning in oncologic histopathology. J. Oral Pathol. Med. **49**(9), 849–856 (2020)

37. Tayebi, A., et al.: Mobile app for comprehensive management of orthodontic patients with fixed appliances. J. Orofacial Orthop./Fortschritte der Kieferorthopädie (2022). https://doi.org/10.1007/s00056-021-00370-7

38. Thomaz, E., Parnami, A., Bidwell, J., Essa, I., Abowd, G.D.: Technological approaches for addressing privacy concerns when recognizing eating behaviors with wearable cameras. In: Proceedings of the 2013 ACM International Joint Conference on Pervasive and Ubiquitous Computing, pp. 739–748 (2013)

39. Tosaka, Y.: Analysis of tooth brushing cycles. Clin. Oral Invest. **18**(8), 2045–2053 (2014). https://doi.org/10.1007/s00784-013-1172-3

40. Underwood, B., Birdsall, J., Kay, E.: The use of a mobile app to motivate evidence-based oral hygiene behaviour. Br. Dent. J. **219**(4), E2–E2 (2015). https://doi.org/10.1038/sj.bdj.2015.660

41. Wang, A., Yu, L., Lan, Y., Zhou, W., et al.: Analysis and low-power hardware implementation of a noise reduction algorithm. In: 2021 International Conference on High Performance Big Data and Intelligent Systems (HPBD&IS), pp. 22–26. IEEE (2021)

42. Wickramasuriya, J., Datt, M., Mehrotra, S., Venkatasubramanian, N.: Privacy protecting data collection in media spaces. In: Proceedings of the 12th Annual ACM International Conference on Multimedia, pp. 48–55 (2004)
43. Wikipedia contributors: Mel-frequency cepstrum. https://en.wikipedia.org/wiki/Mel-frequency_cepstrum. Accessed 13 July 2022

Construction and Evaluation of a Return Prediction Model for One-Way Car Sharing

Ryota Saze[1](✉) , Manato Fujimoto[2,3] , Hirohiko Suwa[1,3] , and Keiichi Yasumoto[1,3]

[1] Nara Institute of Science and Technology, Ikoma, Nara 630-0192, Japan
{saze.ryota.sl4,h-suwa,yasumoto}@is.naist.jp
[2] Osaka Metropolitan University, Osaka 558-8585, Japan
manato@omu.ac.jp
[3] RIKEN, Tokyo, Japan

Abstract. One-way ECS (Electric Car Sharing Service) is attracting attention as a new sustainable mobility option in urban areas. On the other hand, the vehicle uneven distribution problem occurs in one-way ECSs due to their usage patterns. In this paper, we propose a vehicle return prediction model for vehicle relocation to solve this problem. In the proposed method, two machine learning models are created to predict where and when a user will return a vehicle using static information such as departure time and location, and dynamic information such as the vehicle's current location and direction of movement. The model is used to continually update the prediction results of vehicle returns during use, aiming for more accurate predictions. The proposed method has been evaluated using actual data from a one-way ECS and has achieved an accuracy of 0.93 for the prediction of stations to be returned. The method also achieved a MAE of 42.3 min and MAPE of 47% for the prediction of return times.

Keywords: Electric Car Sharing · Machine Learning · Return Prediction · Vehicle Uneven Distribution Problem

1 Introduction

In recent years, one-way electric car sharing services (hereinafter referred to as "one-way ECS") have become popular as a form of mobility in urban areas. In fact, in Japan, TIMES MOBILITY CO., LTD. has launched a service that can be used in Tokyo[1]. In addition to being flexible for short-distance travel in urban areas, one-way ECSs are expected to reduce CO_2 emissions by reducing gasoline consumption [1]. On the other hand, in one-way ECS, there is a "vehicle uneven distribution problem" in which the demand for vehicles is unevenly distributed

---

[1] https://share.timescar.jp/roadway/.

© ICST Institute for Computer Sciences, Social Informatics and Telecommunications Engineering 2023
Published by Springer Nature Switzerland AG 2023. All Rights Reserved
J. Taheri et al. (Eds.): MobiCASE 2022, LNICST 495, pp. 59–71, 2023.
https://doi.org/10.1007/978-3-031-31891-7_5

at certain times and places. If this situation occurs, the system's ability to meet the demand from users will be compromised, resulting in a decline in the rate at which the system meets usage requirements. A solution to this problem is "vehicle relocation" in which vehicles are relocated by the system operator or by the users themselves [6]. In order to properly perform this vehicle relocation, it is necessary to predict the future vehicle deployment and vehicle relocation in advance in such a way as to prevent the occurrence of vehicle uneven distribution problem.

There are some studies that predict vehicle demand for each station, but few studies predict future vehicle deployment by predicting the return station and return time for each vehicle in use.

In this paper, we propose a method for constructing a return prediction model that predicts the return station and return time of vehicles in real time by capturing the movement direction vector of each vehicle during use as a feature, and evaluate the constructed model by using data obtained from a car sharing service that is actually in operation. As a result, it is confirmed that the proposed method improves the return prediction results while the vehicle is moving. The contribution of this study is three folds:

1. First, we propose a return prediction model to predict the return station and return time for a one-way ECS.
2. Second, we evaluate using data obtained from a one-way ECS in real world.
3. Finally, we confirm that the prediction results are improved over time by using real-time vehicle movement direction vectors and other features as explanatory variables.

The rest of this paper is organized as follows. In Sect. 2, we briefly review the existing studies related to our study. We then describe the challenges of a one-way ECS in Sect. 3. Our proposed method for vehicle return prediction in one-way ECS is described in Sect. 4. We then present the evaluation method of our method and the results in Sect. 5. Finally, Sect. 6 concludes this paper.

2   Related Work

One-way ECSs have been the subject of various studies, including comparisons with the conventional round-trip car-sharing method, analysis of user usage characteristics, and research to gain insight into actual operations [3,10,14].

Huang et al. [4] compared two relocation methods, operator-based and user-based. Senju et al. [11] proposed a method of user-based vehicle relocation and confirmed that it improves the utilization demand fulfillment rate in simulations assuming a small-scale one-way ECS. Yang et al. [19] proposed an integrated model for operator-based vehicle relocation to minimize daily operational costs by optimizing request acceptance, relocation tasks, and relocation staff travel routes. Wang et al. [17] proposed a combined user-based and operator-based vehicle relocation method that considers the impact of dynamic constraints such as user demand, vehicle state of charge, and operating profit. Wang et al. [18] also proposed an adaptive co-location model that combines ECS and cycle share to increase ECS utilization. In this research, they achieved a 70% relocation cost reduction compared to staff-based relocation. Luo et al. [9] used multi-agent reinforcement learning to model and treat vehicle relocation tasks, and confirm that demand fulfillment and net benefits are improved using actual operational data.

In order to implement vehicle relocation, some studies have been made to predict the demand from users, and determine where to relocate. Luo et al. [8] extracted features from a graph representing various correlations (distance, point of interest, etc.) between stations of a one-way ECS using multi-graph convolution, and used them to predict the demand for vehicles per station. Yu et al. [20] performed LSTM-based vehicle demand forecasting using temporal features such as time of day, day of week, and weather. Huo et al. [5] used real-time data as input and built a data-driven optimization model that combines a stochastic expectation model and a linear programming problem to deal with the vehicle relocation problem. Wang et al. [16] used application log data of ECS users, historical usage data, as well as real-time station data and user's personal data. Wang et al. [13] proposed a method to improve the operational efficiency of ECS by dynamically changing the frequency of demand forecasting in ECS according to the amount of demand at each time of day.

On the other hand, there are several previous studies on predicting vehicle returns, which is another necessary element for vehicle relocation. Wang et al. [15] constructed a model to predict which stations are likely to return vehicles by calculating trajectories with high similarity by matching real-time GPS trajectory data with historical trajectory data. Liu et al. [7] utilized real-time vehicle trajectory data and application log data in addition to historical usage data to calculate the probability that a user will return a vehicle within 15 min and the probability of returning a vehicle to stations that can be reached within 15 min. The method proposed in their paper calculates the probability that a user will return a vehicle within 15 min and the probability of returning a vehicle to each station that can be reached within 15 min.

However, while there have been studies on vehicle return prediction based on real-time vehicle data, to the best of the authors' knowledge, there have been no studies that predict the return station and return time from the time a user starts using a vehicle, and that continue to improve the prediction accuracy over time by updating the prediction results during the user's use of the vehicle. In this study, we propose a method for predicting vehicle returns at regular intervals using static information such as the start time of use and the starting station, and dynamic information such as the real-time movement direction vector and location of the vehicle.

3 Challenges of a One-Way ECS

3.1 One-Way ECS

A one-way ECS is a type of ECS that allows vehicles to be returned to a station different from where they were rented. As described in Sect. 1, the vehicle uneven distribution problem occurs in one-way ECSs. This is due to the fact

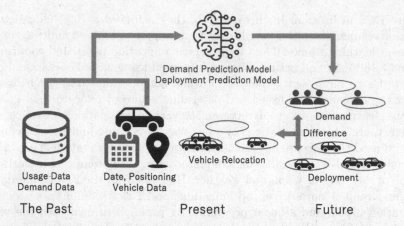

Demand Prediction Model
Deployment Prediction Model

Demand

Difference

Vehicle Relocation

Deployment

Usage Data
Demand Data

Date, Positioning
Vehicle Data

The Past   Present   Future

Fig. 1. Conceptual diagram of vehicle relocation in a one-way ECS.

that the demand for vehicles is unbalanced temporally and spatially, resulting in stations with no vehicles or overflowing with vehicles. There are two ways to solve this uneven distribution problem: operator-based vehicle relocation, in which personnel hired by the system operator relocate vehicles, and user-based vehicle relocation, in which users are requested to reallocate their vehicles.

3.2   Vehicle Relocation

In order to perform appropriate vehicle relocation, it is necessary to predict the future utilization demand and vehicle deployment in the ECS. Figure 1 shows the relationship between vehicle relocation and the prediction of utilization demand and vehicle deployment in a one-way ECS. As shown in the figure, vehicle relocation is possible by predicting both vehicle demand and vehicle deployment.

The demand prediction is necessary to determine in advance which stations will have a high concentration of demand in the future and to determine the necessity of relocation. On the other hand, vehicle deployment prediction is necessary to reduce unnecessary relocation of vehicles in combination with future demand predictions. For example, suppose a situation where a large demand is predicted for a certain station in the future. If the future vehicle deployment is not predicted, it is not known how many vehicles currently in use will be returned to that station, and unnecessary relocation of vehicles may occur. This would result in extra operational costs, and thus reduce the operational efficiency of the ECS.

3.3   Vehicle Deployment Prediction

Vehicle deployment prediction can be achieved by predicting when and where a vehicle in use will be returned to a station based on historical and real-time vehicle usage data. In this paper, we define this as "Vehicle Return Prediction", and in particular, we define the task of predicting where a vehicle in use will be returned as "Return Station Prediction", and the task of predicting when a vehicle will be returned as "Return Time Prediction". Even if the return station prediction can be done with high accuracy, if the time of return is not known accurately, the future deployment of vehicles will be ambiguous. In the opposite case that the place of return is unclear and the time of return is accurate, it will be assumed that the future placement of vehicles will be incorrect. Therefore, both of these prediction tasks must be performed with high accuracy for proper vehicle return prediction.

4   Proposed Method

We assume that for accurate vehicle return prediction, it is necessary to take into account the real-time direction of vehicle movement and the time since the vehicle was first boarded. Thus, we propose a return prediction model that predicts vehicle returns using real-time vehicle data, aiming to realize optimal vehicle relocation for solving the vehicle uneven distribution problem.

4.1   Summary of Proposed Methodology

The proposed method predicts the return of a user's vehicle every 5 min from the start of the user's use of a vehicle. The vehicle return prediction is based on static information such as the user's departure station and departure time, and dynamic information such as the vehicle's location and direction of movement, using machine learning. Note that the return station prediction and return time prediction are performed using different models, and therefore two machine learning models are used in the proposed method.

4.2   Features Used in the Model

Table 1 shows the information used to create the return prediction model. The *Start slot* indicates the time slot in which the user used the vehicle. In this study, one day is divided into 20-min time slots, and 72 slots are assumed per day. The *User ID* is an ID that identifies each user of the ECS. The *Ori station* and *Dst station* are the stations where the user started and finished using the ECS vehicle. In this study, each station is encoded as a number. The *Day of week* is the day of the week when used. We encoded *Day of week* as integers from 0 to 6, e.g., Monday is encoded as 0. The *Elapsed time* indicates the time elapsed since the start of use. The unit of Elapsed time is minutes. The *Latitude* and *Longitude* are the GPS position information at the time the data was saved. The *Movement direction* is the calculated azimuth angle moved from 5 min ago when the north direction is $0[°]$. The *Total usage time* is the total time a user uses the ECS.

Table 1. Variables in return prediction models.

Column	Variable	Unit
Total usage time	Objective	Minutes
Dst station	Objective	Integer
Start slot	Explanatory	Integer
User ID	Explanatory	Integer
Ori station	Explanatory	Integer
Day of week	Explanatory	Integer
Elapsed time	Explanatory	Minutes
Latitude	Explanatory	Fractional
Longitude	Explanatory	Fractional
Movement direction	Explanatory	Fractional

4.3   Return Station Prediction Model

The return station prediction model was created by training XGBoost [2] on a classification task. For the return station prediction, the *Start slot*, *User ID*, *Ori station*, *Day of week*, *Elapsed time*, *Latitude*, *Longitude*, and *Movement direction* in the collected data were set as explanatory variables, and the *Dst station* was set as the objective variable.

4.4   Return Time Prediction Model

The return time prediction model was created by training XGBoost on a regression task. The return station prediction used the same explanatory variables as the return station prediction, but the objective variable was changed to *Total usage time*. It is assumed that the actual time of return is calculated using the total usage time and the start time of usage predicted by the model.

5   Evaluation

The evaluation experiment was conducted using the one-way ECS service data[2] operated by the Nara Institute of Science and Technology (NAIST). We used the data between June 2020 and May 2021 to train two machine learning models of the proposed method and used the data from June 2021 to test the models. The return station prediction model, which is a classification task, was evaluated using the confusion matrix and the F value for each class. The regression task, return time prediction model, was evaluated using Mean Absolute Error (MAE) and Mean Absolute Percent Error (MAPE), which was introduced to correctly evaluate the importance of the same error for long and short total usage.

---

[2] https://naist-carshare.github.io/.

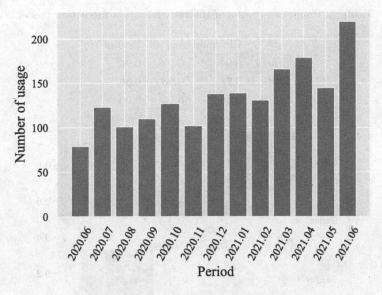

Fig. 2. Usage amount for each month.

Table 2. Number of times used at each original/destination station.

Ori	Dst		
	School	Station	Facility
School	1208	202	28
Station	215	33	13
Facility	16	21	24

5.1 Data Collection

The data for this evaluation was collected from a one-way ECS that is in operation at NAIST. This is operated on a scale of three vehicles and three stations in total operation. The three stations are located in a school, near a railroad station, and near a research facility, respectively, and are referred to as "School", "Station", and "Facility". Each vehicle operated by the ECS is equipped with a GPS module, and GPS positioning data is stored every minute when the vehicle is in operation and available. The ECS is managed with a time slot similar to the one assumed in Sect. 4.1. Users bid for the time slots they wish to use using their own tokens. The user who has bid with the most tokens at the time of the start of the time slot wins the right to use the service. the number of usage for each month is shown in Fig. 2, and the number of times each station is used as an original and a destination is shown in Table 2.

5.2 Evaluation of Return Station Prediction

Prediction Result of Each Stations. Table 3 shows the results of the return station prediction model for the test data, evaluated separately for each station.

Table 3. Prediction results of the return station prediction model.

	Precision	Recall	F-value
School	0.94	0.98	0.96
Station	0.77	0.58	0.66
Facility	0.96	0.84	0.90

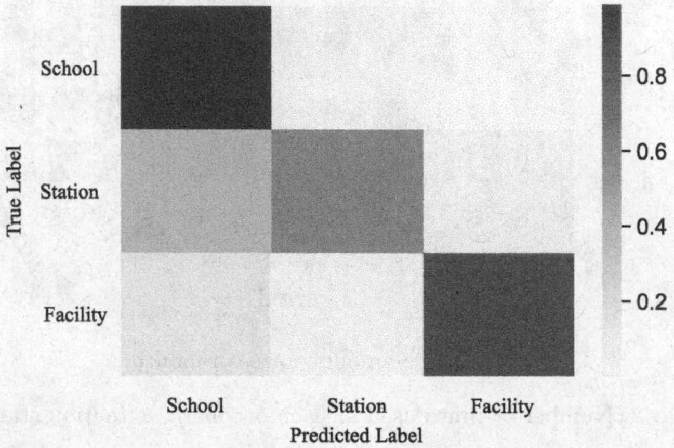

Fig. 3. Confusion matrix of return station prediction model's precision.

This table shows that the station prediction model is able to predict stations with a high degree of accuracy. In fact, the percentage of correct predictions for the entire test data was as high as 0.93. The reason for this high accuracy can be attributed to the large number of usage for School in the test data, and the high accuracy of the prediction for the use where the destination station is a School, as can be seen from the table.

Also, Fig. 3 shows a confusion matrix. Figure shows that the predictions were highly accurate when the target station was a Facility. The reason for this is that only a small number of users used the Facility as their destination station in the usage data, and the high accuracy is thought to be due to the fact that the *User ID* features were used as input. However, when the destination station is the Station, the prediction accuracy is lower than in the other cases. The reason for this is yet to be clarified, and future efforts should be made to improve the accuracy by adding distance information to each station as a feature value.

Time-Series Change of Station Prediction Results. The proposed method updates time-series information, such as the *Movement direction* from the start of use to the end of use, and makes predictions. Intuitively, the closer the end of use is, the closer the predicted return station is to the target station, so the prediction accuracy should improve. In fact, about half of the incorrect predictions, that were made at beginning of the use, actually became correct

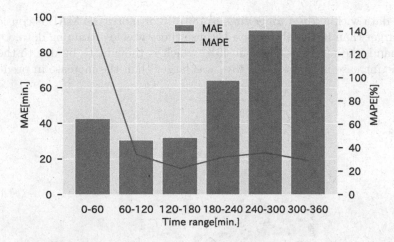

Fig. 4. Results of the return time prediction model for each time-range.

near the end of the use. Therefore, it is possible to improve prediction accuracy by continuing to update features and make predictions in a time-series manner. In the future, it is necessary to confirm what characteristics (travel route, user ID, etc.) are observed in these cases in which the predicted results did not change to correct ones.

5.3 Evaluation of Return Time Prediction

Accuracy of Return Time Prediction. The prediction results of the return time prediction model for all test data were 42.3 min for MAE and 47% for MAPE. Figure 4 shows the results of comparing MAE and MAPE for the usage divided by the *Total usage time*. The bar graph in the figure shows MAE and the line graph shows MAPE. From the results of Fig. 4, MAPE was high at 146% for use within 60 min. This is a reasonable value considering that the MAE is 42.6 min, but considering the application to a real system, the value needs further improvement.

On the other hand, for the use between 60 and 180 min, the MAE was about 31 min in both cases, and the MAPE was 34% and 22% lower than in the previous time period. In June 2021, the median of the total time of use[3] was 1.6 h, and in nearly half of the cases, the *Total usage time* was between 60 and 180 min. This indicates that a practical model for predicting the return time has been obtained.

---

[3] https://naist-carshare.github.io/logs/2021-07-02-log-202106/.

For data with a *Total usage time* of 180 min or more, the MAE is considered to be larger than in the above case because there was less training data. On the other hand, the MAPE was about 40%, which is thought to be due to the fact that the increase in *Total usage time* was larger than the increase in prediction error.

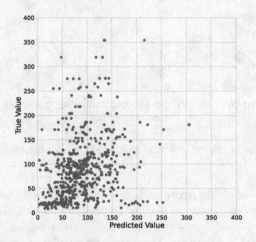

Fig. 5. Result for within 15 min from the end of use. This result shows the model's *Total usage time* prediction results for the input data within 15 min of the start.

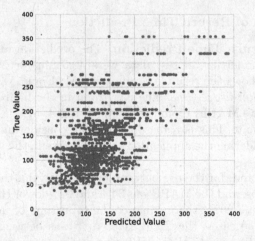

Fig. 6. Result for 15 min after the start - 15 min before the end. This result shows the model's *Total usage time* prediction results for the input data from 15 min after the start to 15 min before the end.

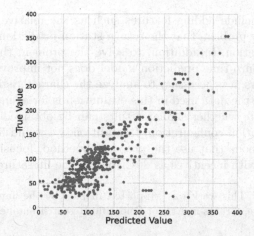

Fig. 7. Result for within 15 min from the end of use. This result shows the model's *Total usage time* prediction results for the input data within 15 min of the end.

Time-Series Changes in Time Prediction Results. Figures 5, 6, 7 shows the prediction results of the return time prediction model within 15 min of the user's start of use, from 15 min after the start to 15 min before the end, and within 15 min of the user's end of use. The closer to the diagonal line from the lower left to the upper right of the graph, the more accurate the prediction results are. In the Fig. 5, the model predict shorter use times than the true value in many cases. Figure 6 show an increase in density around the diagonal compared to Fig. 5. Furthermore, Fig. 7 were clearly more concentrated around the diagonal than those for the earlier time periods. This indicates that the prediction accuracy of the actual return time improves with the *Elapsed time* of use. Therefore, it is shown that the proposed method improves prediction accuracy over time. On the other hand, there were cases where large prediction errors occurred even within the last 15 min of use, so it is necessary to improve the return time prediction model by analyzing the characteristics common to these uses.

6   Conclusion

In this study, we propose a vehicle return prediction model that is necessary for vehicle relocation to solve the vehicle uneven distribution problem in a one-way ECS. The proposed method performs vehicle return prediction by adding real-time vehicle location and direction information in addition to the user's past usage history, and keeps updating the prediction results while the vehicle is in use. The evaluation results showed that the proposed method achieved a high accuracy rate of 0.93 for the entire test data in the return station prediction. MAE and MAPE were smaller for the 60 to 180 min time period than for the earlier time period. Furthermore, we confirmed that the prediction results became more accurate as the time of use increased.

Future plans include adding features such as the relative distance to each station to solve the problem that the return station prediction model is inaccurate for specific station. In addition, to solve the problem that the prediction accuracy of the return time prediction model does not improve even at the end of use for some users, it is necessary to analyze the characteristics of these usage. Furthermore, since the data used in this evaluation is for a small ECS, it is necessary to investigate whether similar results can be obtained for a larger ECS with a different station configuration. In addition, we would like to study vehicle relocation methods that use the proposed method, focusing on user-based vehicle relocation with incentives as described in the literature [12].

Acknowledgment. This work was supported in part by the Japan Society for the Promotion of Science, Grants-in-Aid for Scientific Research number JP21H03431 and JP21K11879.

References

1. Baptista, P., Melo, S., Rolim, C.: Energy, environmental and mobility impacts of car-sharing systems. Empirical results from Lisbon. Portugal. Proc. Soc. Behav. Sci. **111**, 28–37 (2014)
2. Chen, T., Guestrin, C.: XGBoost: a scalable tree boosting system. In: Proceedings of the 22nd ACM SIGKDD International Conference on Knowledge Discovery and Data Mining, pp. 785–794 (2016)
3. Curtale, R., Liao, F., van der Waerden, P.: Understanding travel preferences for user-based relocation strategies of one-way electric car-sharing services. Transp. Res. Part C: Emerg. Technol. **127**, 103135 (2021)
4. Huang, K., An, K., Rich, J., Ma, W.: Vehicle relocation in one-way station-based electric carsharing systems: a comparative study of operator-based and user-based methods. Transp. Res. Part E: Logist. Transp. Res. **142**, 102081 (2020)
5. Huo, X., Wu, X., Li, M., Zheng, N., Yu, G.: The allocation problem of electric car-sharing system: a data-driven approach. Transp. Res. Part D: Transp. Environ. **78**, 102192 (2020)
6. Jorge, D., Correia, G.H., Barnhart, C.: Comparing optimal relocation operations with simulated relocation policies in one-way carsharing systems. IEEE Trans. Intell. Transp. Syst. **15**(4), 1667–1675 (2014)
7. Liu, D., Lu, J., Ma, W.: Real-time return demand prediction based on multisource data of one-way carsharing systems. J. Adv. Transp. **2021** (2021)
8. Luo, M., Du, B., Klemmer, K., Zhu, H., Ferhatosmanoglu, H., Wen, H.: D3P: data-driven demand prediction for fast expanding electric vehicle sharing systems. Proc. ACM Interact. Mob. Wearable Ubiquit. Technol. **4**(1), 1–21 (2020)
9. Luo, M., et al.: Rebalancing expanding EV sharing systems with deep reinforcement learning. In: Proceedings of the Twenty-Ninth International Conference on International Joint Conferences on Artificial Intelligence, pp. 1338–1344 (2021)
10. Park, S., Yu, W.: Analysis of system parameters for one-way carsharing systems. Transp. Lett. 1–11 (2021)
11. Senju, K., Mizumoto, T., Suwa, H., Arakawa, Y., Yasumoto, K.: Designing strategy for resolving maldistribution of vehicles in one-way car-sharing through active

trip request to potential users. In: 2018 IEEE International Conference on Pervasive Computing and Communications Workshops (PerCom Workshops), pp. 83–88. IEEE (2018)

12. Stokkink, P., Geroliminis, N.: Predictive user-based relocation through incentives in one-way car-sharing systems. Transp. Res. Part B: Methodol. **149**, 230–249 (2021)

13. Wang, G., Qin, Z., Wang, S., Sun, H., Dong, Z., Zhang, D.: Record: joint real-time repositioning and charging for electric carsharing with dynamic deadlines. In: Proceedings of the 27th ACM SIGKDD Conference on Knowledge Discovery & Data Mining, pp. 3660–3669 (2021)

14. Wang, G., Vaish, H.R., Sun, H., Wu, J., Wang, S., Zhang, D.: Understanding user behavior in car sharing services through the lens of mobility: mixing qualitative and quantitative studies. Proc. ACM Interact. Mob. Wearable Ubiquit. Technol. **4**(4), 1–30 (2020)

15. Wang, L., Zhong, Y., Ma, W.: GPS-data-driven dynamic destination prediction for on-demand one-way carsharing system. IET Intel. Transport Syst. **12**(10), 1291–1299 (2018)

16. Wang, L., Zhong, H., Ma, W., Zhong, Y., Wang, L.: Multi-source data-driven prediction for the dynamic pickup demand of one-way carsharing systems. Transportmet. B: Transp. Dyn. **8**(1), 90–107 (2020)

17. Wang, N., Guo, J., Liu, X., Liang, Y.: Electric vehicle car-sharing optimization relocation model combining user relocation and staff relocation. Transp. Lett. **13**(4), 315–326 (2021)

18. Wang, N., Liu, Q., Guo, J., Fang, T.: A user-based adaptive joint relocation model combining electric car-sharing and bicycle-sharing. Transportmet. B: Transp. Dyn. **10**(1), 1046–1069 (2022)

19. Yang, S., Wu, J., Sun, H., Qu, Y., Li, T.: Double-balanced relocation optimization of one-way car-sharing system with real-time requests. Transp. Res. Part C: Emerg. Technol. **125**, 103071 (2021)

20. Yu, D., Li, Z., Zhong, Q., Ai, Y., Chen, W.: Demand management of station-based car sharing system based on deep learning forecasting. J. Adv. Transp. **2020** (2020)

Dependable Systems

A Bi-directional Attribute Synchronization Mechanism for Access Control in IoT Environments

Bruno Cremonezi[1], Luciano F. da Rocha[2], Alex B. Vieira[2], José Nacif[3], André L. de Oliveira[2], and Edelberto Franco Silva[2]([⊠]) [iD]

[1] Federal University of Paraná - UFPR, Curitiba, Brazil
[2] Federal University of Juiz de Fora University - UFJF, Juiz de Fora, MG, Brazil
edelberto@ice.ufjf.br
[3] Federal University of Viçosa - UFV, Viçosa, Brazil
http://edelbertofranco.ice.ufjf.br

Abstract. The Attribute-Based Access Control (ABAC) model is widely used for IoT due to its capacity to express access policies through attributes, making this method granular and flexible. However, if we assume that attributes are essentially mutable, the irreducible network latency and the architectures proposed to acquire a better communication performance of the IoT expose the point where those policies are evaluated as outdated attributes. Therefore, access policies can be wrongly evaluated, resulting in consistency and security problems. In this paper, we propose a method to reduce this exposure through a bi-directional attribute synchronization capable of mapping all attributes and evaluating their current consistency after a change. If the modified attribute does not affect the access, it will remain valid. Otherwise, a revocation occurs, reducing the risks of unintended accesses. Our modeling allows demonstrating the correctness of our method and its capability to revoke every unintended access that may occur after an attribute change.

Keywords: IoT · Access Control · ABAC · Age of Information · UPPAAL

1 Introduction

The Internet of Things (IoT) is a technological trend in which common everyday objects are now equipped with sensing and communication capabilities. Therefore, the so-called IoT devices are becoming increasingly popular in our lives and today form a hyper-connected ecosystem of devices, enabling the emergence of several revolutionary applications on the market [9]. Although the benefits of this hyper-connected ecosystem to society with the provision of applications that enable automation, convenience, and effectiveness for everyday tasks, it also raises several concerns regarding the security and privacy of its users [18].

IoT devices are present in all sectors of our lives, collecting, accessing, and transferring information, often confidential or critical. Therefore, ensuring

© ICST Institute for Computer Sciences, Social Informatics and Telecommunications Engineering 2023
Published by Springer Nature Switzerland AG 2023. All Rights Reserved
J. Taheri et al. (Eds.): MobiCASE 2022, LNICST 495, pp. 75–88, 2023.
https://doi.org/10.1007/978-3-031-31891-7_6

that such information is not transferred to malicious locations and/or individuals, authentication and access control are essential and critical tasks for IoT devices [18]. Concerning access control, the attribute-based model (ABAC) is widely adopted in IoT applications due to its flexibility and expressiveness [16]. In an access control model, access decisions are taken based on identification attributes assigned to people, objects, or environments present in a hyperconnected IoT ecosystem against a previously defined access policy [6]. However, attributes and policies are mutable, and ideally, the entity that eváluates the access policy, i.e., the decision point, should have all these values in real-time. In practice, keeping all these values in real-time at the decision point is unfeasible. Because, even though the attributes and access policies can be accessed in real-time, there is an irreducible latency of the network that introduces a risk that some values arrive outdated at the decision point [12]. Moreover, in order to improve authorization performance, by avoiding sending attribute and policy query requests over the network, attributes and policies are usually stored in *caches*, whose values could be outdated when requested. Therefore, due to the irreducible latency of the network or outdated *caches*, some access decisions taken may be incorrect, leading to security and consistency problems [12].

In this paper, we investigate the problem of security and consistency in an IoT environment concerning the attributes of users and objects. As a first step to address this problem, we propose an attribute mapping model, a bi-directional synchronization model for all attributes, and a model capable of determining the consistent state of the attributes in the network to determine whether the access remains valid or must be revoked after an attribute update. Here, we used two approaches: if the access remains valid, an access update is performed, keeping its validity for the previously established time window; in case the access becomes invalid, its revocation occurs immediately. It is important to highlight although the consistency problem is extensively explored in practice, this work explores the formal modeling of this problem through timed automata. We seek to demonstrate the correctness of the proposal and its ability to deliver correct accesses at the end of a given execution.

The remainder of this work is organized as follows: Sect. 2 presents related works. Section 3 describes the bi-directional attribute synchronization method. Section 4 details the evaluation methodology. Section 5 discusses the results obtained. Finally, Sect. 6 presents conclusions and future directions.

2   Related Work

The security and consistency problem considered in this work can be addressed through a quick update of the attributes at the decision point. Within this scope, there is a number of studies addressing the issue of consistency in distributed systems, ranging from classic [1,5] to contemporary [11,15] methods. As highlighted by [17], many access control models are not fully compatible with the assumptions of distributed systems, being directed to more static environments or using reactive queries to attributes and policies.

Considering ABAC as an access control method, we have that it has premises of distributed environments due to its flexibility and granularity. Thus, we keep the focus on this access control model and its related work regarding consistency definition and credential updates. The previous related work closest to the studied concept is by *Lee and Winslett* (LW) [13,14]. Although the work of [17] is inspired by the previous one, a new perspective is considered since it evaluates the updating and not the revocation of a policy. On the other hand, our proposal evaluates the updating and consistency of attributes for use in ABAC applied to IoT and computational *fog*. In this paper, we propose the evolution and creation of a new research topic related to the investigation introduced by [17]. Our work is one of the first to evolve the concept of update and consistency operation for attributes in a distributed access policy scenario.

3   Assumptions and System Modeling

This work operates under the *eXtensible Access Control Markup Language* (XACML) standard. We chose XACML by being a consolidated authorization standard and explicitly defined for the access control model ABAC [3]. XACML offers specifications that cover all ways of using ABAC, from policy definitions to architecture, to support this model. Moreover, several related works available in the literature point to the XACML model as suitable to be used in an IoT scenario [7]. To illustrate it, Fig. 1 presents a diagram with the entities specified by the XACML model and the order of messages exchanged between them. According to the XACML model, four entities are needed to implement ABAC: *Policy Enforcement Point* (PEP), *Policy Decision Point* (PDP), *Policy Information Point* (PIP), and the *Policy Administration Point* (PAP). In this work, access policies are considered immutable. However, it is important to note that the entity responsible for updating and distributing such policies to the PDPs is PAP (0) [16].

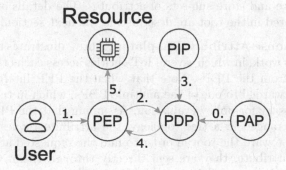

Fig. 1. XACML architecture.

In Fig. 1, when a given IoT user wants to access a given object, e.g., performing a reading operation on a device, the PEP intercepts this request (1) and generates an authorization request which is forwarded to the PDP (2). However, to assess whether or not this access should be authorized, the PDP looks for an access policy previously stored through the PAP and queries which attributes are needed to evaluate this policy through the PIP (3). Based on the policies and attributes, the PDP evaluates the access and sends its decision to the PEP (4), which allows or not the access of the user (5) [16].

Users, Attributes and Policies: In this work, for convenience, the term user is used to determine something/someone that requests access to a certain computational resource. However, it is important to note that a user could be a person, service, application, or even another IoT device. Regardless of the nature of the user, it is assumed that he/she has one or more identities with a set of attributes that describe them. Usually, in large applications - and even in medium-sized applications - it is common for the user to have several identities that represent him/her stored into several different locations [10]. The attributes of identities are mutable and vary between discrete values. Regarding access policies, this work uses immutable policies divided into rules. Each rule presents a set of attributes and the values they must have to determine if an access is valid or not. If any policy rule is satisfied, the access is valid. Otherwise, if all rules are not satisfied, the access is considered invalid [12].

PIP, PAP, and PDP: This work operates under an ABAC authorization environment with multiple authorities. It is assumed that there are multiple PDPs, PEPs, and PIPs distributed over a large geographic area, which serve access requests from various IoT entities. This scenario is quite common in many applications to improve the performance of the authorization process [16]. Additionally, this work determines that the PIP is segmented in a network arranged in a tree topology. The root node represents an extensive repository of attributes capable of storing the attributes of all IoT entities, and the nodes below, in turn, represent the attribute caches commonly used to obtain better authorization performance and store sub-sets of attributes. The details of the operation of this PIP located in the root are described in the next section.

Request-Response Attribute Template: Figure 2 illustrates the architecture proposed in this work, in which several IoT entities access each other and request authorizations from the PDPs. Note that when the PEP intercepts an access request, it is forwarded to one of the multiple PDPs, which in turn requests the attributes necessary for policy evaluation. Assuming that the PIP is segmented in a network arranged in a tree topology, the attribute request goes through multiple caches toward the root in order to find the requested attributes. If any cache has the attribute, they are sent directly through them. Otherwise, this request reaches the root of the tree that responds to attributes with a high latency [8].

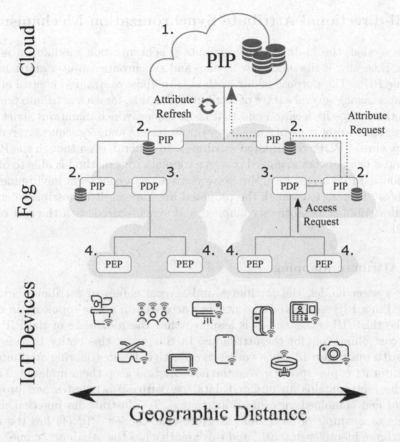

Fig. 2. Architecture of the authorization system

3.1   Problem Formulation

Managing people's attributes, IoT devices, and applications is an essential and challenging task as attributes can change frequently. For example, suppose a user accesses information from IoT devices related to multiple academic projects. This user can change position, join new projects, change location, etc. Similarly, new devices can be added to projects, other devices can be removed, and many other changes can happen. All these changes, although seemingly minor, can pose a significant challenge when using multiple PIPs to store identities. Assuming the ABAC operates in an environment with multiple authorities and several PIPs, for every access that occurs, an attribute query must be performed to determine whether the access is valid or not. However, as attributes are changeable over time and the PIP is segmented into a network arranged in a tree topology, all attributes must be updated in all caches consistently in the tree. Or, if access was previously wrongfully allowed, it must be revoked. Therefore, our main objective is to limit PDP exposure to outdated attributes, update them consistently and ensure that current and future accesses occur securely.

4   Bi-directional Attribute Synchronization Mechanism

In this section, the bi-directional attribute synchronization mechanism is presented. Basically, it discusses how to map and synchronize all user attributes in multiple PIPs. The purpose of our method is to allow centralized control of user attributes among several *caches* of attributes. For this, for each attribute present, the method maps its location and that of its copies, which maintains strict control over the current state of consistency of the system and, consequently, reduces the exposure of PDPs to outdated attributes. In general, even though the PIP is segmented in a network arranged in a tree topology, our method is able to offer a synchronized PIP across its entire network to allow a consistent environment of attributes. Therefore, through the proposed mechanism, it is possible to ensure that all attributes are correctly mapped and synchronized with the rest of the network.

4.1   Attribute Mapping

In our system model, the attributes and several copies of attributes are distributed in a PIP segmented in a network arranged in a tree topology. In order to make this PIP consistent, it is assumed that the root node of the PIP tree has a controlling role for the attributes. In this work, this entity is named as **attribute manager**. In other words, its objective, besides offering attributes, is to monitor its copies spread across the network and keep them updated. Therefore, the root contains an updated database with all attributes and provides a global and combined view of all attributes. To illustrate its function, Fig. 3 presents an example of its usefulness. Note that the left PIP (1) has the identity with the identifier "user01" and this identity has the attribute "name" with the value "Alex". Similarly, the PIP on the right (2) also has the identity with the identifier "user01", however, instead of the "name", this identity has the attribute "position" with the value "Professor". As many PIPs do not have the full view of users and attributes may be missing in others, the attribute manager allows an overview of the user. Therefore, the attribute manager has the attributes of both identities and creates a complete identity with the attributes "name" and "position". In addition, the attribute manager creates a base that points out in which PIPs the identity is stored. If this identity is removed from *cache* PIPs or added to others, these PIPs must send an attribute map update message to the manager.

4.2   Attribute Synchronization

As attributes are mutable, any PIP can perform an attribute update operation. However, to reflect this change in other PIPs, it is needed to perform attribute synchronization. In this work, we propose the mechanism of the bi-directional synchronization of attributes. It has this name because, at first, it is sent to the attribute manager (update occurs "up"), and the manager, from its attribute map, updates all *caches* (update occurs down). To exemplify this process, Fig. 4

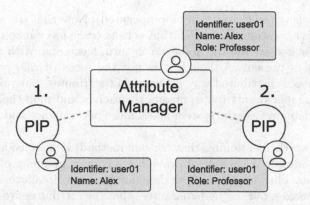

Fig. 3. Attribute Mapping

illustrates an attribute update. Assume that user "user01" has its identity repli-cated in both PIPs (1 and 2). Assume that, due to some operational change, this user's title changed from "teacher" to "researcher" and this change occurred in PIP 2. (a). PIP 2. forwards this change to the attribute manager (b), which updates the global view of the system (c). After updating the global view, it searches its attribute map in which PIPs this attribute was stored and updates them (d).

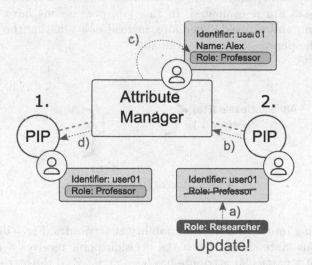

Fig. 4. Attribute Synchronization

It is important to note that the synchronization process takes place through an essential operation. There is a PIP, which is the source of the change (i.e., the PIP where an attribute was changed), and a PIP, which is the target of the change (PIP where the change will be propagated). Note that the map serves as a guide for this operation. Every PIP that acts as *cache* has a map that points to the attribute manager, and the manager, in turn, has a map with all the *caches* that attribute is present. Although it is not the focus of this paper, and we will not address a distributed way of managing attributes, our proposal allows PIPs *cache* to replicate attributes among themselves and map their replicas for a possible update, in case the system needs more performance and a distributed spread.

It is also worth mentioning that, in our method, it is considered that the **attribute manager** is capable of providing a correct mapping of the location of all attributes. Therefore, as long as there are no network failures or difficulties in sending messages, our method guarantees that the attributes are synchronized in the *cache* PIPs and, consequently, in the decision points for decision-making and access revalidation. However, there is no guarantee that these attributes were not modified or attacked in the *cache* PIPs. However, this scenario is considered outside the scope of this work.

4.3 Access Revalidation

In this work, it is assumed that when a PDP makes an access decision, it maintains a history of this permission (P), which shows which access rule was met and, consequently, which attributes were used to make the decision ($P = a1, a2, ...$). Therefore, after an attribute change, the PIP announces a change to the PDP, and the accesses are re-evaluated. In this job, permissions have three states: valid, unknown, and revoked. Therefore, instead of evaluating the entire access policy, only the permission is first evaluated.

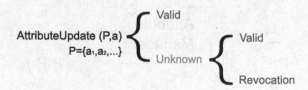

Fig. 5. Access Revalidation

When taking into account, the mutability of attributes, Fig. 5 illustrates how the permissions state change when the decision point receives a message that informs it that a particular attribute has been updated. If the attribute does not change permission, i.e., the modified attribute is not used in that permission, the access permission will remain **valid**. If the attribute is used, but the result of that access rule's decision does not change, the permission will also remain **valid**. If the attribute affects the access permission in such a way that the rule that granted it permission becomes invalid, this access will not become invalid, but **unknown**. For these cases, the entire access policy must be re-evaluated through the new attribute. If any other rule of the access policy is satisfied, this access permission returns to the **valid** state and is updated with the new rule that satisfies it. If the access no longer satisfies any other rule, an access **revocation** occurs, which is immediately communicated to the PEP, which suspends the user's access to the resource.

5 Results

For the evaluation of the proposed bi-directional synchronization mechanism of attributes, we carried out a formal verification through timed automata. Conceptually, timed automata is a generalization of finite automata to a continuous-time domain. In addition to the traditional transitions and states, the automaton also has a finite number of real variables, called clocks, whose values increase with derivative 1 concerning the passage of time. Each automaton transition can be constrained by clock values and can only occur if a particular condition is satisfied. In general, no clock modification operations exist except for the reset operation. Moreover, it is also important to note that the value of a clock can only be compared against rational constants and not against the value of other clocks [2].

5.1 The UPPAAL Tool

The UPPAAL [4] tool allows system modeling by defining several basic automata in an editor. In addition, the tool has a system trajectory simulator and an automatic property verification module. The verification module uses algorithms and data structures available in the literature to perform a *model-checking* of the system, which is a Cartesian product of basic automata, against properties expressed in a subset of TCTL logic (*Timed Computation Tree Logic*).

In TCTL logic, conceptually, the quantifier A denotes "for every trajectory", while the quantifier E denotes "there is a trajectory". For analysis, these quantifiers must be combined with the quantifiers <> and [], which denote, respectively, "in some state of the trajectory (*eventually*)" and "in all states of the trajectory". Therefore, the UPPAAL tool can present in its simulator an example and a counter-example of a given expression φ when a property of type $E <> \varphi$ is true (example) or when a property of type $A[]\varphi$ is false (counter-example). For the present work, we only used the basic properties of reachability expressed by the predicate $E <> \varphi$, which denotes the existence of a trajectory in which the formula φ becomes valid at some future time.

5.2 Scenario and Models

This work considers an IoT application in which a user requests access to several devices - such as thermostats, for example - in different locations [17]. Consider the scenario in Fig. 6 for illustrative purposes. Suppose Alex is a newly hired teacher and is located near PEP (1). Your identity has 3 attributes: Role, Trust, and Location. At an instant of time t_1, he started working, and his identity was pegged with the lowest trust level, for example, the value 1. In this application, an access policy denies users all operations on the IoT device if its trust level is 1. After some time, a top Alex user updates his trust level to the value 2 on the base PIP (1). At that moment, Alex performs access at the time instant t_2 and is allowed, for example, to turn the device on/off. Now suppose Alex changes his location to PEP(2). If the update level attribute has been synchronized with the PIP (2), Alex will be able to perform access at the instant t_3. Otherwise, the access will be denied incorrectly. This work enumerates the conditions for this access to be incorrectly denied. This situation is modeled below.

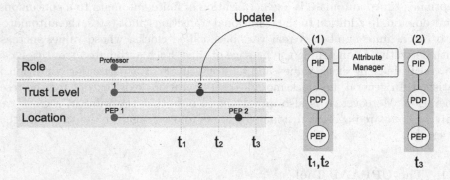

Fig. 6. Evaluation scenario

The automaton shown in Fig. 7 models the scenario shown in Fig. 6. This automaton controls user behavior and offers the action of requesting access (CLIENT_PLEASE_REQUEST_ACCESS), updating the attributes of an identity (PIP_PLEASE_CHANGE_ATTRIBUTES) and changing the location by the of a user (CLIENT_PLEASE_CHANGE_LOCATION). It is important to note that the "!" and "?" are synchronism between the automata. Simply put, a transition with the "?" implies a transition waiting for synchronization, whereas a transition with the operator "!" activates synchronization.

For all communication that occurs between users and XACML entities, we model sending (_SENT) and receiving messages (_RECEIVED) through an automaton that models the communication channel (Fig. 8). For every message sent (_SENT), the communication channel implies a communication delay that causes a specific time instant t to pass. Only then does the other entity receive it (_RECEIVED). For example, in our channel, when an access request is sent to the PEP, the channel synchronizes this message (ACCESS_REQUEST_SENT[e]?), waits an instant of time t, and only then forwards this message to the PEP via the synchronization message (ACCESS_REQUEST_RECEIVED[channel]!). Note that all messages shown in Fig. 1 are modeled in this channel through ACCESS_REQUEST, DECISION_REQUEST, ATTRIBUTE_REQUEST, ACCESS_RESPONSE, DECISION_RESPONSE, ATTRIBUTE_RESPONSE synchronizations. Furthermore,

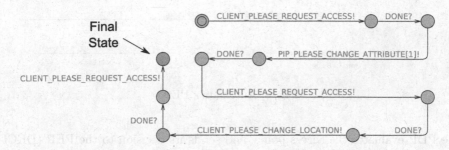

Fig. 7. Automate - Evaluation Scenario

the synchronization of updated attributes in a given PIP with the attribute manager is modeled in the GA_SYNC_ATTRIBUTE message.

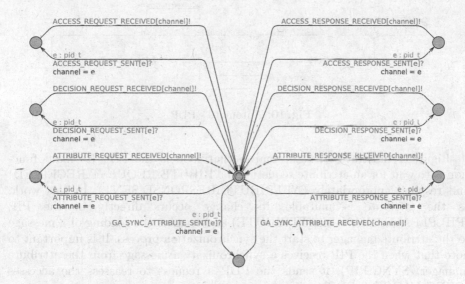

Fig. 8. Automate - Communication channel

Figure 9 illustrates the automaton that models the PEP. When intercepting a client access request (ACCESS_REQUEST_RECEIVED), the automaton sends a decision request (DECISION_REQUEST_SENT) to the PDP and waits for the PDP to respond. After responding (DECISION_RESPONSE_RECEIVED), the PEP interprets the PDP response and sends it to the client (ACCESS_RESPONSE_SENT).

Figure 10 illustrates the automaton that models the PDP. Upon receiving a decision request from the PEP (DECISION_REQUEST_RECEIVED), the automaton sends to the PIP an attribute request (ATTRIBUTE_REQUEST_SENT) and waits for the PIP to respond. After the PIP sends the requested attributes (ATTRIBUTE_RESPONSE_RECEIVED),

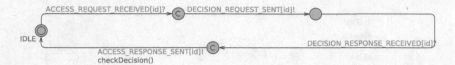

Fig. 9. Automata - PEP

the PDP evaluates the access policy and sends its decision to the PEP (DECI-SION_RESPONSE_SENT). It is important to note that, throughout its operation, the PDP can be requested to re-evaluate access after changing a specific attribute, updating unaffected accesses, and revoking incorrect accesses (CHECK_ACCESS_PDP) as presented in Sect. 4.3.

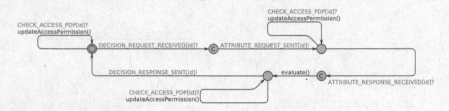

Fig. 10. Automate - PDP

Figure 11 illustrates the automaton that models the PIP. Its main function is to wait for an attribute request (ATTRIBUTE_REQUEST_RECEIVED) and respond appropriately (ATTRIBUTE_RESPONSE_SENT). In this work, as the attribute is mutable, its change occurs directly in the PIP (PIP_PLEASE_CHANGE_ATTRIBUTE) and triggers the sending of a message to the attribute manager to start the synchronization process. It is important to note that when the PIP receives a synchronization message from the attribute manager (SYNC_PIP), it sends the PDP a request to reassess the accesses (CHECK_ACCESS_PDP).

Finally, to close the models, Fig. 12 presents the automaton that represents the attribute manager. In general, its main function is to wait for attribute synchronization requests (GA_SYNC_ATTRIBUTE_RECEIVED) from a PIP and send the request to the other PIPs where the attribute is mapped (SYNC_PIP).

5.3   Formal Verification of Models

To start the formal verification of our model, we define three basic properties to be achieved. In the first moment, we want to verify if the model is free of stops and correct. Therefore, the following expression was used: $E <> deadlock$. In our tests, the time limit was 30 min, and we could not find any unexpected stops. While this does not prove that the model is free of *deadlocks*, it indicates that it is correct. Our second expression was used to check the existence of

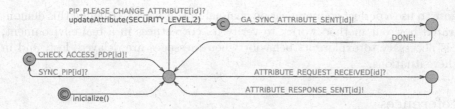

Fig. 11. Automate - PIP

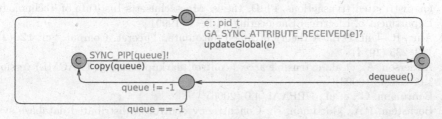

Fig. 12. Automate - Attribute Manager

improper accesses at the instant $t3$. For this, the following expression was used: $E <> PEP[2] \cdot invalidAuthorization()$. Translating it: "Is there a trajectory where there is invalid access in PEP 2?". In this case, the simulator pointed it out as true and presented several examples in which if the synchronization process occurs after the access request in PEP 2, this access is granted incorrectly. Therefore, to verify that our method revokes these accesses correctly, we use the third expression: $E <> PEP[2] \cdot invalidAuthorization() \&\& SCENARIO \cdot END$. In this case, the simulator pointed out this expression as false. As much as there exist accesses that may have been granted improperly, at one point, our method is capable of re-validate them, and, at the end of the verification, there is no invalid access. Thus, we demonstrated that our method, at some point, can recognize improper access and revoke it correctly.

6   Conclusion

This work presented a study of real-time aspects of the bi-directional synchronization method of attributes through its modeling and verification in the UPPAAL tool. We used the Timed Automata formalism and assumed a perfect channel model to carry out a complete verification of the protocol that allowed us to identify situations and conditions for improper access. However, at the same time, our verification demonstrated that all unauthorized accesses at the end of the simulation were revoked. The results point to two directions for future work. On the one hand, it would be interesting to implement a distributed version of our method since, as much as our centralized approach removes the consistency problem, it can lead to scalability and fault tolerance problems. Moreover, we chose to use a perfect communication channel in this work. Although this demonstrated how our method works, to verify its correctness in a real environment, it is necessary to explore its behavior when messages are delayed, lost, and in other situations.

References

1. Adya, A.: Weak consistency: a generalized theory and optimistic implementations for distributed transactions. Ph.D. thesis, Massachusetts Institute of Technology, Department of Electrical Engineering and . . . (1999)
2. Alur, R., Dill, D.L.: A theory of timed automata. Theoret. Comput. Sci. **126**(2), 183–235 (1994)
3. Anderson, A., et al.: extensible access control markup language (XACML) version 1.0. OASIS (2003)
4. Behrmann, G., et al.: UPPAAL 4.0 (2006)
5. Bernstein, P.A., Goodman, N.: Concurrency control in distributed database systems. ACM Comput. Surv. (CSUR) **13**(2), 185–221 (1981)
6. Bezawada, B., Haefner, K., Ray, I.: Securing home IoT environments with attribute-based access control. In: Proceedings of the Third ACM Workshop on Attribute-Based Access Control, pp. 43–53 (2018)
7. Caserio, C., Lonetti, F., Marchetti, E.: A formal validation approach for XACML 3.0 access control policy. Sensors **22**(8), 2984 (2022)
8. Cremonezi, B., Gomes Filho, A.R., Silva, E.F., Nacif, J.A.M., Vieira, A.B., Nogueira, M.: Improving the attribute retrieval on ABAC using opportunistic caches for fog-based IoT networks. Comput. Netw. **213**, 109000 (2022)
9. Dian, F.J., Vahidnia, R., Rahmati, A.: Wearables and the internet of things (IoT), applications, opportunities, and challenges: a survey. IEEE Access **8**, 69200–69211 (2020)
10. Garbis, Jason, Chapman, Jerry W..: Identity and access management. In: Garbis, J., Chapman, J.W. (eds.) Zero Trust Security, pp. 71–91. Springer, Heidelberg (2021). https://doi.org/10.1007/978-1-4842-6702-8_5
11. Harding, R., Van Aken, D., Pavlo, A., Stonebraker, M.: An evaluation of distributed concurrency control. Proc. VLDB Endow. **10**(5), 553–564 (2017)
12. Hu, V.C., et al.: Guide to attribute based access control (ABAC) definition and considerations (draft). NIST Spec. Publ. **800**(162), 1–54 (2013)
13. Lee, A.J., Winslett, M.: Safety and consistency in policy-based authorization systems. In: Proceedings of the 13th ACM Conference on Computer and Communications Security, pp. 124–133 (2006)
14. Lee, A.J., Winslett, M.: Enforcing safety and consistency constraints in policy-based authorization systems. ACM Trans. Inf. Syst. Secur. (TISSEC) **12**(2), 1–33 (2008)
15. Perrin, M.: Distributed Systems: Concurrency and Consistency. Elsevier, Amsterdam (2017)
16. Ravidas, S., Lekidis, A., Paci, F., Zannone, N.: Access control in internet-of-things: a survey. J. Netw. Comput. Appl. **144**, 79–101 (2019)
17. Shakarami, M.: Operation and administration of access control in IoT environments. Ph.D. thesis, The University of Texas at San Antonio (2022)
18. Tawalbeh, L., Muheidat, F., Tawalbeh, M., Quwaider, M., et al.: IoT privacy and security: challenges and solutions. Appl. Sci. **10**(12), 4102 (2020)

Fault Tolerance in Cloud Manufacturing: An Overview

Auday Al-Dulaimy[1,2]([✉]) [iD], Mohammad Ashjaei[1] [iD], Moris Behnam[1] [iD], Thomas Nolte[1] [iD], and Alessandro V. Papadopoulos[1] [iD]

[1] Mälardalen University, Västerås, Sweden
{auday.aldulaimy,mohammad.ashjaei,moris.behnam,thomas.nolte,
alessandro.papadopoulos}@mdu.se
[2] Dalarna University, Falun, Sweden
aay@du.se

Abstract. Utilizing edge and cloud computing to empower the profitability of manufacturing is drastically increasing in modern industries. As a result of that, several challenges have raised over the years that essentially require urgent attention. Among these, coping with different faults in edge and cloud computing and recovering from permanent and temporary faults became prominent issues to be solved. In this paper, we focus on the challenges of applying fault tolerance techniques on edge and cloud computing in the context of manufacturing and we investigate the current state of the proposed approaches by categorizing them into several groups. Moreover, we identify critical gaps in the research domain as open research directions.

Keywords: Cloud computing · Edge computing · Cloud manufacturing · CMfg · Manufacturing as a Service · MaaS · Fault tolerance · Reliability

1 Introduction

To drive profitability, manufacturers adopted the concept of Cloud Manufacturing (CMfg) in their business, as it offers manufacturing services with lower cost and better performance. CMfg builds on top of cloud and edge computing, and it uses the infrastructure on cloud data centers (cloud layer) and on the factory-site computing servers (edge layer) to transform traditional manufacturing resources into services.

Cloud computing has huge computing and storage capabilities, however, its classical centralized architecture comes with some limitations [1]. These limitations include (1) A stable connectivity between factory sites and cloud data centers is required to offer convenient services, (2) Cloud computing assumes that

---

This research was partially sponsored by the Knowledge Foundation (KKS) under the SACSys project, and by the research center XPRES (Excellence in Production Research) - a strategic research area in Sweden.

J. Taheri et al. (Eds.): MobiCASE 2022, LNICST 495, pp. 89–101, 2023.
https://doi.org/10.1007/978-3-031-31891-7_7

there is enough bandwidth to transfer data between the manufacturers' devices and the cloud data centers, (3) Massive data transfer causes network bottlenecks leading to latency and performance deterioration issues, and (4) Data transfer results in security concerns. On the other hand, edge computing tries to overcome such limitations by providing decentralized and closer to factory-site computing and storage resources. However, such resources usually have limited capacities [1, 8].

In CMfg, the above-mentioned limitations in cloud and/or edge layers may result in system failures. A system failure can be defined as [12]: *"An event in which the system fails to operate according to its specifications. A system failure occurs, when a system deviates from fulfilling its normal system function for which it was aimed at."* Besides the limitations accompanied by cloud and edge layers, the production phases that consist of dynamic and long life-cycle processes add more complexity to the systems [26], and such complexity comes with more difficulties in providing reliable and fault tolerances services in the manufacturing environments. Therefore, there is a need to manage the failures that may occur in the services offered to the manufacturing and industrial sectors. Without proper fault tolerance approaches, multiple manufacturing services will fail to lead to great losses. A better understanding of the topics related to fault tolerance in the context of manufacturing will help improve productivity and increase profitability. Therefore, the aim of this work is to cover the essential directions related to system failures in CMfg.

1.1 Contributions

The main contributions of this paper are summarized as follows:

- investigating the system failures in the context of the CMfg environment (specifically the edge and cloud layers), with a focus on failures at the edge layer, and presenting an overview of the issues related to system failures.
- presenting the research gaps that are associated with the existing fault tolerance approaches in the edge layer. Such gaps need to be investigated when designing new fault tolerance systems, taking the features of the edge devices into consideration, which are also listed in this paper.

1.2 Paper Layout

The rest of this paper is organized as follows: Sect. 2 presents the taxonomy of the paper. It presents a background on reliability and fault tolerance in cloud and edge computing. In addition, it gives a brief literature review of the existing fault tolerance approaches and categories them. Section 3 defines the research gaps in the topic. Finally, Sect. 4 concludes the paper.

2 Taxonomy

The paper presents examples of cloud manufacturing in practice, followed by a detailed discussion on the reliability and fault tolerance issues in the context of a cloud manufacturing environment. The detailed discussion covers the following perspectives:

- Cloud manufacturing in practice
- Features of the edge devices
- Reliability in edge-cloud environments
- Reasons for system failures in edge-cloud environments
- Fault tolerance solutions in edge-cloud environments

The work in this paper is focusing on the system failures in the *edge layer*. Several existing works (*e.g.*, [18, 24], and [22]) survey fault tolerance and its related topics in cloud computing environments, not specifically looking at the edge. Finally, complementary to the work presented in this paper, the work in [3] compares and shows the differences/similarities between the edge and cloud layers.

2.1 Cloud Manufacturing in Practice

Recently, manufacturing started broaden its overall objective from production-oriented to include more cases of service-oriented manufacturing, and as a consequence the cloud service providers began to offer what is called Manufacturing as a Service (MaaS). With MaaS the manufacturers can outsource the services and new industrial technologies related to all process stages to a trusted party, while concentrating on the innovation and core mission. Among the main pioneers in providing MaaS, we can mention the following:

Google Cloud for Manufacturing. Google cloud aims at helping manufacturers to transform into a digital environment by providing innovative solutions that reshape the production and factory-floor operations [10].

Amazon Cloud for Manufacturing. Amazon provides advanced digital transformation solutions to manufacturers. Such solutions utilize machine learning and data analysis to optimize production and improve operational efficiencies [4].

Microsoft Cloud for Manufacturing. Microsoft offers manufacturing services that drive productivity and improve security. The core processes and requirements of the industry are encapsulated and provided as capabilities from Microsoft aiming at providing secure connection and resilient business processes. Such capabilities enhance the time-to-value metrics for manufacturers in a scalable fashion. Microsoft cloud and edge resources are integrated with smart components for providing different manufacturing scenarios such that the beneficiaries select the highest-value scenario [16].

However, to provide stable manufacturing services, the providers need to offer and maintain fault tolerant systems at the edge and cloud layers.

2.2 Features of the Edge Devices

The infrastructure at the edge computing layer has specific characteristics. In particular, the edge devices are featured by the following:

– Constrained devices, *i.e.*, they have limited compute power and fixed storage capacity.
– Geo-distributed devices.
– Heterogeneous devices.
– Connectivity cannot be guaranteed with the cloud layer.
– Edge devices run containers more efficiently compared with Virtual Machines (VMs).

The devices at the edge of the network need to perform communication and computation tasks in order to provide real-time responses for a large number of end devices at the Manufacturing layer. They are connected horizontally with each other, and vertically across layers (either upwards with the cloud layer or downwards with the manufacturing layer) [7].

The aforementioned characteristics and design goals must be considered when proposing fault tolerance approaches in the edge layer environment.

2.3 Reliability in Edge-Cloud Environment

Reliability in service-oriented edge-cloud computing, which is adopted and used by manufacturers, is how consistently the services are provided without interruption and/or failure. A failure is a state when any system fails to operate according to its design goals, or when the system can not work according to a specific predefined Quality-of-Service (QoS). Fault tolerance is a way to prevent or deal with failures, such that the system continues operating and providing services regardless of the failure type. In CMfg, fault tolerance approaches are essential to meet the manufacturers' requirements, and to understand the infrastructure needed to provide persistent manufacturing services.

Designing a robust fault tolerant system in CMfg requires a deep understanding of the reasons and types of failures, and how the systems should respond to such failures. This will be discussed in the following sections.

2.4 Reasons for System Failures

Understanding the causes of the occurrence of the failure is fundamental in proposing the appropriate solution to avoid or deal with failures. The reasons for failures in a manufacturing environment, which comprises the edge and cloud layers, can be categorized into five main categories, as shown in Fig. 1.

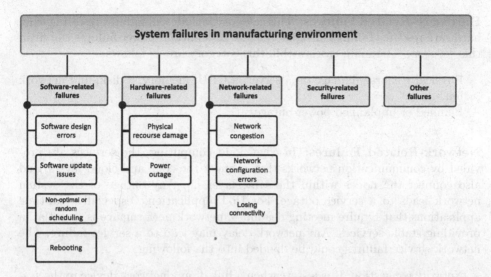

Fig. 1. The reasons of system failures in manufacturing environments.

Software-Related Failures: This category includes all failures resulting from the running software systems and applications. Recently, the software is getting complex and accompanied by diverse functional and non-functional requirements. Moreover, the software is more sophisticated as they are designed to work in an edge-cloud environment. Therefore, the software became a significant reason for system breakdown which causes loss in business and revenue. Failures may occur due to the following situations in this category:

- Software design errors: Designing software in the right way is essential in avoiding many errors which may lead to system faults [28]. The common fault examples include implementing incorrect infinite loops, numerical overflow/underflow, and no protection against deadlocks.
- Software update issues: As a result of security issues, or to enrich the applications with additional features, the software is regularly updated with new batches. This can make the software volatile to faults.
- Non-optimal or Random scheduling: Usually, the designed software has to be executed together with other applications with either data or time dependencies. This normally requires proper scheduling for the software to avoid any blocking or dependency issues among the software applications. A naive scheduling method can increase the chance of faults during the run-time of the software.
- Rebooting: The software can be rebooted, either planned or unplanned, during the execution of the system, which can potentially stop serving the running applications, and bring new faults that were not covered before.

Hardware-Related Failures: This category includes all failures resulting from hardware resource failures or replacements. In addition, power failures can fit in this category. The main cases within this category are as follows:

- Physical recourse damage on a server, a CPU, memory, a disk, and network links.
- Planned or unplanned power outage.

Network-Related Failures: In edge-cloud computing, the services are provided by communication networks that connect the edge and cloud layers, and also connect the nodes within the same layer [2]. Any outage of the system network leads to a service outage. For IIoT applications, especially real-time applications that require meeting deadlines, network performance is essential in providing stable services. Any network delay may lead to a service failure. The network service failures could be divided into the following:

- Network congestion: It is a state when a link or any network device in the system is forwarding a huge amount of data which can result in over-consumption of the communication bandwidth. This can violate the QoS requirements but is also considered a fault in the network.
- Network configuration errors: It covers the processes of assigning network settings, policies, controls, and data flows [30]. In the edge-cloud environment, the design and infrastructure of the networking are virtualized and then implemented by underlying software across physical network devices. The proper network configuration is essential in supporting the network flow and stability, otherwise, it may lead to network failures.
- Losing connectivity: Availability of the network is a metric that is also dictated by a Service Level Agreement (SLA) and violating that can affect the overall system performance.

Security-Related Failures: Several security issues lead to failures, such as viruses and malicious. In an edge-cloud environment, the system must be able to defend against malicious attacks and provide a trusted storage and computing base, otherwise, the security attacks may lead to a system failure.

Other Faults: In this category, we can include many examples that are not categorized in the previous groups, such as human errors or natural disasters (*e.g.*, earthquakes).

2.5   Dealing with Failures

The approaches to deal with failures in the manufacturing environment can be classified into four different categories, as depicted in Fig. 2. In the following, these categories are elaborated with more details.

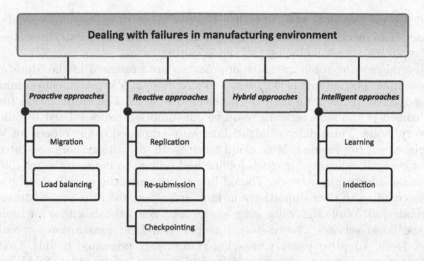

Fig. 2. Dealing with failures.

Proactive: To maintain reliability, some cloud service providers adopt proactive approaches to avoid possible failures before their occurrences. This way is used to predict the faults proactively and substitute the suspected component with some running components. The proactive approaches mainly can be sub-classified into two classes [27]: migration and load balancing.

Several existing robust proactive approaches are presented in the literature. For example: In [20], the authors proposed an approach that aims at preventing system faults within the federated cloud environments. The environment is modeled as a multi-objective optimization problem that maximizes the profit and minimizes the VMs migration cost. The approach redistributes VMs from the expected faulty providers (based on the CPU temperature) to healthy ones within the federation. The work in [29] proposes an approach to predict preemptive migration decisions using a Generative Adversarial Network (GAN) in a containerized edge environment. The proposed model, called *PreGAN*, detects and classifies faults to schedule migrations to obviate the potential system faults. In [27], the authors proposed an approach, called Automated Pipeline for Advanced Fault Tolerance (APAFT), to monitor the task production rates at the edge layer and check the ability of the edge computational nodes to serve these tasks. APAFT predicts the potential bottlenecks in task execution that may result in potential system faults, and accordingly triggers proactive node replication. In [17], the authors compare different proactive approaches based on control theory and probabilistic model checking for the autoscaling of cloud-based systems.

Reactive: This way is mainly used to decrease the influence of failure and provide reliability to the system after the failures have occurred. Reactive approaches take some measures in order to react accordingly. In general, such approaches may result in a large overhead and expensive implementation, but cloud service providers need to utilize reactive ways to manage any potential failures. The reactive approaches mainly can be sub-classified into three classes [27]: replication, re-submission, and checkpointing.

Several existing robust reactive approaches are presented in the literature. For example: The authors in [33] presented a redundant VM placement optimization approach to secure a fault-tolerant cloud system. The optimization function considers the huge network resource consumption issues related to failure recovery mode. Three different algorithms were employed in the process of VM (re)placement to provide reliable cloud services. In [31], the authors presented a two-stage fault tolerance approach (offline and online) to improve the reliability of the manufacturing network. The off-line stage ranks the manufacturing services according to their importance in fault tolerance, then the critical services are replicated. While the online stage performs a heuristic algorithm for replacing the failed services. A two-stage unsupervised fault recognition approach, called Deep Adaptive Fuzzy Clustering (DAFC), is presented in [11]. DAFC integrates two clustering algorithms, Stacked Sparse AutoEncoder (SSAE) and Adaptive Weighted Gath-Geva (AWGG), aiming at proposing an unsupervised fault recognition framework to cluster unlabeled big data in the manufacturing environment and then extract fault features from the clusters. In [15, 19], the authors analyze the usage of control-based reactive load balancing techniques for masking potential faults in the data center.

Hybrid: To maintain the maximum possible level of reliability and availability in cloud manufacturing, several approaches consider both proactive and reactive ways to deal with system failure.

In the literature, very few hybrid approaches have been proposed. For example, in [5], the authors presented a hybrid model to take fault tolerance actions, including proactive actions after predicting the failure probability, and reactive actions that employ replication and checkpointing techniques. The work in [24] presented a fault-tolerance approach that utilizes two directions. The first direction is to perform VM migrations based on a failure prediction technique, while the second direction is to implement the checkpointing process. Moreover, the work in [1] presented an approach, called TOLERANCER, that aims to solve the software and hardware-related failures in a cloud manufacturing environment. TOLERANCER composes of connected components that are collaborating with each other to detect stress situations or node failures, and accordingly, trigger actions to avoid and solve potential system failures. The work presented in [25] describes a hybrid checkpointing mechanism implemented in OpenStack, that can be used for optimizing the usage of resources based on the incoming workload while improving the fault-tolerance capacity of the virtualized environment.

Intelligent: Such approaches try to handle application requirements when faults happen and, at the same time, improve the service within an appropriate time frame. Intelligent systems are resilient and include smart elements that are able to deal with the application requirements during any disruptions aiming at reaching a system's safe status. Such approaches share common features with the proactive approaches, such as monitoring the system and predicting faults to avoid them, but they differ from proactive approaches in utilizing intelligent learning [18]. The Intelligent approaches can be sub-classified into two classes [18,22]: learning and induction.

Selected intelligent approaches can be mentioned: In [6] the authors stated that fault diagnosis is essential in offering stable services in industry. Aiming at identifying and preventing system failures, they used simple vibration data and applied different unsupervised learning algorithms that test the performance and robustness of the system. The work in [32] utilized deep learning to propose a fault identification approach in industrial systems. To diagnose faults, the approach extracts features from the system, and then a classification model is used to detect fault information.

In [23] the authors stated that most of the existing load balancing algorithms in cloud environments ignore fault tolerance in their design. Thus, they presented an algorithm that employed fault tolerance metrics in a load balancing approach. The approach works in three phases: *Observing phase* to collect system information aiming at identifying any disorganized behaviors, *Inspection phase* to specify the relation or differentiate between defects aiming at false diagnosis, and *Organize and Implementation phase* to reassign a correct weight to the damaged elements aiming at storing healthy system state.

3   Findings and Research Gaps

Studying the fault tolerance approaches presented in Sect. 2.5, and exploring the state of the art, *e.g.*, [13,14], led us to define the following issues to be tackled when designing new fault tolerance approaches in the edge-cloud layers.

- In the context of cloud manufacturing, manufacturers are utilizing both the edge and the cloud layers. Hence, the designed fault tolerance approach must be holistic in considering different aspects, e.g., long-distance network connectivity (vertical network) between different layers, network connectivity (horizontal network) within the same layer, security and privacy issues, and all related failures discussed in Sect. 2.4.
- Edge and cloud resources have different features, as discussed in Sect. 2.2. Thus, the designed fault tolerance approaches need to have separate implementations for the edge and the cloud layers.
- IIoT applications are commonly real-time applications that have certain timing requirements. Therefore, the decisions of the designed fault tolerance approaches need to be made at run-time to avoid failures.
- The cloud manufacturing environment is dynamic and serves different IIoT applications which are accompanied by diverse functional and non-functional requirements. Thus, fault tolerance approaches in such environments need to be smart and able to learn and adapt to the system environment. The use of learning and induction algorithms seems promising in providing solid fault-tolerant systems for manufacturers.

– Hybrid approaches combine the advantages of both proactive and reactive approaches, hence it is always better to be adopted compared with considering proactive or reactive approaches individually.
– There are many technical issues in the existing fault tolerance approaches to tackle in order to meet their design goals. Among many, we can identify the following technical issues:

1. Checkpointing [9] in the reactive approaches performs efficiently at the edge layer as the resulting checkpoint files have small sizes which can be stored on the edge nodes of fixed storage capacity. However, it is possible to further speed up the checkpointing process, which results in faster service retrieval and continuing the ongoing application(s).

2. Function as a Service (FaaS) [21] is a promising platform under the umbrella of cloud computing as it allows developers to run their applications without considering the complexities related to building or maintaining the infrastructure. In cloud manufacturing, manufacturers could get benefit from FaaS if it is used at the edge layers. However, there are crucial drawbacks with FaaS at the edge layer which need to be solved. For instance, hosting long-running function instances on constrained edge devices may not be feasible due to the memory requirements of Dockers which deliver software in packages called containers. In addition, the computation cannot be paused and continued later (stateless). Therefore, novel methods to manage function containers are needed to overcome these drawbacks.

3. Theoretically, integrating FaaS with checkpointing looks promising in providing solid fault-tolerant systems. However, some challenges may appear with this combination. For instance, how to specify the sleep and active timing for the function containers? Proposing visible solutions to such challenges is needed, for example, injecting the system through an external monitoring process to examine file descriptors and incoming network connections.

4   Conclusion and Future Directives

Cloud Manufacturing (CMfg) utilizes the resources at the edge and cloud layers, aiming at providing holistic manufacturing services to maximize manufacturers' profits. Such services should be stable and fault tolerant. Many fault tolerance approaches have been proposed in the literature, however, there are challenges that must be addressed and solved in order to attain reliable and fault tolerant systems in the context of CMfg.

This work presented an overview of the fault tolerance-related issues in CMfg environments, along with the research gaps in such environments. As future directives, we are working on: (1) expanding this work to include more topics related to system failures in CMfg, and (2) solving some of the identified technical issues using hybrid and/or smart approaches.

References

1. Al-Dulaimy, A., Christian, S., Papadopoulos, A.V., Galletta, A., Villari, M., Ash-jaei, M.: Tolerancer: a fault tolerance approach for cloud manufacturing environments. In: IEEE 27th International Conference on Emerging Technologies and Factory Automation (ETFA 2022) (2022)
2. Al-Dulaimy, A., Itani, W., Taheri, J., Shamseddine, M.: bwslicer: a bandwidth slicing framework for cloud data centers. Futur. Gener. Comput. Syst. **112**, 767–784 (2020)
3. Al-Dulaimy, A., Sharma, Y., Khan, M.G., Taheri, J.: Introduction to edge computing. Edge Comput. Models Technol. Appl. 3–25 (2020)
4. Amazon: manufacturing: simplifying digital transformation. https://aws.amazon.com/manufacturing/ (2022). Accessed 2022
5. Amoon, M.: A framework for providing a hybrid fault tolerance in cloud computing. In: 2015 Science and Information Conference (SAI), pp. 844–849. IEEE (2015)
6. Amruthnath, N., Gupta, T.: A research study on unsupervised machine learning algorithms for early fault detection in predictive maintenance. In: 2018 5th International Conference on Industrial Engineering and Applications (ICIEA), pp. 355–361. IEEE (2018)
7. Bakhshi, Z., Rodriguez-Navas, G., Hansson, H.: Dependable fog computing: a systematic literature review. In: 2019 45th Euromicro Conference on Software Engineering and Advanced Applications (SEAA), pp. 395–403. IEEE (2019)
8. Du, W., et al.: Fault-tolerating edge computing with server redundancy based on a variant of group degree centrality. In: Kafeza, E., Benatallah, B., Martinelli, F., Hacid, H., Bouguettaya, A., Motahari, H. (eds.) ICSOC 2020. LNCS, vol. 12571, pp. 198–214. Springer, Cham (2020). https://doi.org/10.1007/978-3-030-65310-1_16
9. Egwutuoha, I.P., Levy, D., Selic, B., Chen, S.: A survey of fault tolerance mechanisms and checkpoint/restart implementations for high performance computing systems. J. Supercomput. **65**(3), 1302–1326 (2013). https://doi.org/10.1007/s11227-013-0884-0
10. Google: Google cloud for manufacturing. https://cloud.google.com/solutions/manufacturing/ (2022). Accessed 2022
11. Hu, X., Li, Y., Jia, L., Qiu, M.: A novel two-stage unsupervised fault recognition framework combining feature extraction and fuzzy clustering for collaborative AIoT. IEEE Trans. Industr. Inf. **18**(2), 1291–1300 (2021)
12. Javadi, B., Thulasiraman, P., Buyya, R.: Enhancing performance of failure-prone clusters by adaptive provisioning of cloud resources. J. Supercomput. **63**(2), 467–489 (2013)
13. Javed, A., Heljanko, K., Buda, A., Främling, K.: Cefiot: a fault-tolerant IoT architecture for edge and cloud. In: 2018 IEEE 4th World Forum on Internet of Things (WF-IoT), pp. 813–818. IEEE (2018)
14. Karhula, P., Janak, J., Schulzrinne, H.: Checkpointing and migration of IoT edge functions. In: Proceedings of the 2nd International Workshop on Edge Systems, Analytics and Networking, pp. 60–65 (2019)
15. Klein, C., et al.: Improving cloud service resilience using brownout-aware load-balancing. In: IEEE 33rd International Symposium on Reliable Distributed Systems (SRDS), pp. 31–40. IEEE, New York (2014). https://doi.org/10.1109/SRDS.2014.14

16. Microsoft: Introducing microsoft cloud for manufacturing. https://www.vmware.com/topics/glossary/content/network-configuration.html (2022). Accessed 2022
17. Moreno, G.A., Papadopoulos, A.V., Angelopoulos, K., Cámara, J., Schmerl, B.: Comparing model-based predictive approaches to self-adaptation: Cobra and PLA. In: 12th International Symposium on Software Engineering for Adaptive and Self-Managing Systems (SEAMS), pp. 42–53 (2017). https://doi.org/10.1109/SEAMS.2017.2
18. Mukwevho, M.A., Celik, T.: Toward a smart cloud: a review of fault-tolerance methods in cloud systems. IEEE Trans. Serv. Comput. 14(2), 589–605 (2018)
19. Papadopoulos, A.V., et al.: Control-based load-balancing techniques: analysis and performance evaluation via a randomized optimization approach. Control. Eng. Pract. 52, 24–34 (2016). https://doi.org/10.1016/j.conengprac.2016.03.020
20. Ray, B., Saha, A., Khatua, S., Roy, S.: Proactive fault-tolerance technique to enhance reliability of cloud service in cloud federation environment. IEEE Trans. Cloud Comput. 10(2), 957–971 (2020)
21. Scheuner, J., Leitner, P.: Function-as-a-service performance evaluation: a multivocal literature review. J. Syst. Softw. 170, 110708 (2020)
22. Shahid, M.A., Islam, N., Alam, M.M., Mazliham, M., Musa, S.: Towards resilient method: an exhaustive survey of fault tolerance methods in the cloud computing environment. Comput. Sci. Rev. 40, 100398 (2021)
23. Shahid, M.A., Islam, N., Alam, M.M., Su'ud, M.M., Musa, S.: A comprehensive study of load balancing approaches in the cloud computing environment and a novel fault tolerance approach. IEEE Access 8, 130500–130526 (2020)
24. Sharma, Y., Si, W., Sun, D., Javadi, B.: Failure-aware energy-efficient VM consolidation in cloud computing systems. Futur. Gener. Comput. Syst. 94, 620–633 (2019)
25. Souza, A., Papadopoulos, A.V., Tomás Bolivar, L., Gilbert, D., Tordsson, J.: Hybrid adaptive checkpointing for virtual machine fault tolerance. In: IEEE International Conference on Cloud Engineering (IC2E), pp. 12–22 (2018). https://doi.org/10.1109/IC2E.2018.00023
26. Tao, F., Zhang, L., Liu, Y., Cheng, Y., Wang, L., Xu, X.: Manufacturing service management in cloud manufacturing: overview and future research directions. J. Manufact. Sci. Eng. 137(4) (2015)
27. Theodoropoulos, T., Makris, A., Violos, J., Tserpes, K.: An automated pipeline for advanced fault tolerance in edge computing infrastructures. In: Proceedings of the 2nd Workshop on Flexible Resource and Application Management on the Edge, pp. 19–24 (2022)
28. Thieme, C.A., Mosleh, A., Utne, I.B., Hegde, J.: Incorporating software failure in risk analysis-part 1: software functional failure mode classification. Reliab. Eng. Syst. Saf. 197, 106803 (2020)
29. Tuli, S., Casale, G., Jennings, N.R.: Pregan: preemptive migration prediction network for proactive fault-tolerant edge computing. In: IEEE INFOCOM 2022-IEEE Conference on Computer Communications, pp. 670–679. IEEE (2022)
30. vmWARE: What is network configuration. https://www.microsoft.com/en-us/industry/manufacturing/microsoft-cloud-for-manufacturing (2022). Accessed 2022
31. Wu, Y., Peng, G., Wang, H., Zhang, H.: A two-stage fault tolerance method for large-scale manufacturing network. IEEE Access 7, 81574–81592 (2019)

32. Xing, D., Chen, R., Qi, L., Zhao, J., Wang, Y.: Multi-source fault identification based on combined deep learning. In: MATEC Web of Conferences, vol. 309, p. 03037. EDP Sciences (2020)
33. Zhou, A., et al.: Cloud service reliability enhancement via virtual machine placement optimization. IEEE Trans. Serv. Comput. **10**(6), 902–913 (2016)

Emerging Applications

Investigating the Adoption of Mobile Government from Citizen's Perspectives in Saudi Arabia

Muneer Nusir[✉] [iD]

Department of Information Systems, College of Computer Engineering and Sciences, Prince Sattam Bin Abdulaziz University, Al-Kharj 16278, Saudi Arabia
m.nusir@psau.edu.sa

Abstract. With the rapidly evolving mobile technology, governments are delivering services to the citizen through a mobile platform. These services include administrative services, health services, and awareness campaigns. To effectively provide mobile services to citizens, it is necessary to understand user perceptions of these services thoroughly. Therefore, the acceptance rate is influenced by a variety of factors. These elements are categorized as social, technological, cultural, personal, or facilitating. This paper aims to present a study on the acceptance of the Mobile-Government (M-Government) system in Saudi Arabia. One of the primary goals of this research is to promote M-Government adoption in developing countries such as Saudi Arabia. As a result, a study is being carried out to determine 'How citizens' cultures and attitudes affect the acceptability of M-Government?' By identifying and analyzing cultural influences on M-Government, it is possible to understand people's needs better. The primary aim of this research is to identify the limitations and research gaps in previous studies and broaden the scope of technology acceptance models to determine the acceptance rate of M-Government services. The Technology Acceptance Model (TAM) and the Unified Theory of Acceptance and Use of Technology (UTAUT) were used to investigate the impact of various factors on M-Government system acceptance. Previous studies' limitations, which are addressed in this paper, include more appropriate constructs added to the models for hypothesis building. These hypotheses are based on Saudi Arabia's demographic profiles, sociological and technological foundations. The findings will help policymakers, and government officials better understand the factors that influence service's user acceptance.

Keywords: M-Government · Saudi Arabia · TAM · UTAUT · Hypothesis Testing · User Acceptance

1 Introduction

The rapid improvements in Information Technology (IT) have revolutionized every aspect of human life. Beginning with conventional services and progressing to technological devices in government employment, the evolution of public services began.

J. Taheri et al. (Eds.): MobiCASE 2022, LNICST 495, pp. 105–116, 2023.
https://doi.org/10.1007/978-3-031-31891-7_8

M-Governments in industrialized nations have found a means to provide information to the public quickly and efficiently. Mobile phones have been in commercial use since 1984 [1]. Several governments have started offering mobile services to residents to improve efficiency and effectiveness. Mobile technology allows access to government services and information anytime and anywhere [2]. Globally, over 7 billion mobile cellular users at the end of 2015, up from 738 million in 2000. Fastest-growing IT ever. The primary goal of M-government is to ensure that the public, businesses, and government are all mobile via the use of wireless devices. As a result of the M-government initiative, residents may save time and energy by using their mobile phones and other wireless devices to access the Internet and government networks.

This research study contributed significantly to the field of IT and will be beneficial to understand better the needs, evaluations, and perceptions of individuals about M-Government services, in both in terms of theory and application. Firstly, when it comes to M-Government in Saudi Arabia, it used a model that hasn't been well studied: the TAM model. Secondly, it is also included some constructs: Facilitating Conditions that enable the acceptance of technology and social influence, to the TAM model, to understand the effect of numerous variables on Saudi residents' behavioral intentions and acceptance of M-Government. Thirdly, Saudi politicians and technology service providers must also better comprehend service delivery from the standpoint of Saudi residents to better serve their constituents in the Kingdom. Fourth, researchers in Saudi Arabia have looked at whether or not M-Government services are well-liked in the country. This study tried to contribute to knowledge in this area by examining if the influence of increased TAM elements on M-Government may be tempered by gender, age, experience, and voluntariness of Saudi residents. Finally, by expanding the breadth of the technology acceptance models, and tried to understand better how people feel about these services. Additionally, efforts were made to resolve the drawbacks of prior research.

Prior research [4–6] on M-Government adoption in countries with comparable cultural, economic, and geographic characteristics to Saudi Arabia is included in the literature review. In this section, many models of technology adoption are examined critically. The methodology section follows the literature review and details the steps taken to access Saudi Arabia's M-Government system. A series of hypotheses were created based on TAM and UTAUT models' unique constructions. These hypotheses are used to build a questionnaire for data collection. Then comes the analysis of the data, which follows. The outcomes of the hypothesis testing were discussed, and the study is ended in the discussion and conclusion parts. The study's shortcomings and potential solutions are discussed in the last detail to guide future research.

2 Literature Review

M-Governments have been an integral part of governance in most countries for several years. Services provided by M-Governments are being consistently improved. The mode of delivery is also improving due to M-Government, being one of the most researched areas in recent years. This section will discuss the contributions made by researchers in the M-Government system and analyze them regarding Saudi Arabia [5, 6]. Despite the availability of M-Government services in countries like Saudi Arabia, they are affected by the low acceptance rate by citizens. In addition, various factors hinder the adoption of these services.

Alonazi et al. [5] have used the M-Government adoption and usability Model 'UTAUT' to study the low usability of M-Government services in the Saudi Arabia [4]. Several parameters like perceived usefulness, social influence, perceived interest, and system quality were assessed in this research. It was concluded that the personal and professional experience with technology was one of the most important elements in the acceptance rate of M-Government services. The study was reported to be a lighthouse for Saudi IT policymakers and Information & communication technology (ICT) professionals. Althunibat et al. [6] have presented a framework of M-Government acceptance. In this research, the advanced statistical methods are used to draw conclusive results from the surveyed data.

The acceptance of M-Government services in Jordan is represented by Althunibat et al. [6]. This survey found that SMS is the most popular service among people. It led to M-Government services in Jordan. The TAM was used to assess M-Government services. The TAM is the most often used model [6–9]. TAM's major components are perceived usefulness and perceived of ease of use. Jordanians strongly seek M-Government services [4, 10, 11] have presented the study on M-Government services in the UAE. An implementation framework based on the Fuzzy Delphi technique was suggested in this research. In the UAE, E-Government gave way to M-Government in 2013. IT skills, IT and Cyber Security Policy, and IT infrastructure were among the challenges faced in the UAE. Saxena et al. used TAM and UTAUT to study the influence of socioeconomic and demographic constraints on M-Government in India [11, 12].

Many researchers have used the term M-Government; however, this study will focus on studies undertaken in Saudi Arabia. M-Government is a strategy and execution that employs mobile technology to improve the benefits of E-government for all parties involved, including people and businesses [13]. According to Joseph and Boateng [14], advances in e-business and M-business have helped governments embrace mobile technology. According to Kim et al. [15], M-Government could give extra information and services such as civic statistics, microfinance, and electronic identity. Finally, Mandari and Koloseni [16] claim that the high penetration of mobility and reachability among people can accomplish E-government via M-Government. Mobile phone use in Saudi Arabia has grown tremendously in the last few years [11, 15]. As a result, M-government services supplied by the Saudi Arabian government are the focus of this study.

Over the past several years, mobile wallets have continuously been in demand and improved, particularly during the COVID-19 pandemic [17]. Mobile communication technology is primarily responsible for the transmission of information, including factors such as information channels' capacity, transmission efficiency, and information security, among others [4, 11]. Therefore, organizations responsible for information management should focus on acquiring, integrating, and expressing information [10, 18]. According to Althunibat et al. [13] was conducted as study, technologies that aid in implementing M-government may be divided into four broad categories:

a) Government-to-Government (G2G) transactions: The use of ICT to conduct commerce inside and between government entities.

b) Transactions between the government and the citizen are known as G2C transactions, and they include the use of ICT to provide m-government services to the public.
c) Transactions between the government and businesses (G2B): The use of ICT to provide M-government services to the business sector.
d) Transactions between government and employees (G2E): The use of ICT to provide and conduce workshops and training among government agencies.

M-government may bring prospective advantages for the public sector, apart from the obstacles, as shown in Table 1. Apart from the issues listed in Table 1, M-Government can provide advantages to the public sector.

Table 1. Benefits and challenges.

Benefits	Challenges
It is boosting the efficiency of government services	Privacy & security
The government served a vast number of people	People Awareness
A public service's efficiency	Trust
Low-priced	Additional costs
Government information and services are made available	Information overload
Use whenever you want	Struggle & Hard-work

In most of the research on the adoption, acceptance, and utilization of M-Government services, social, demographic, or design factors are considered as the literature review has suggested that there are several state affairs, social background, demographic facts, and technical factors that play a decisive role in the acceptance rate of M-Government services. In this research paper, all these factors will be addressed for the Kingdom of Saudi Arabia. In this paper, the researcher used the UTAUT models to build the hypothesis.

2.1 Factors Affecting the Adoption/Acceptance of M-Government

This section discusses the application of technology acceptance models for assessing the acceptance of M-Government in Saudi Arabia. As various researchers have accessed technology acceptance and usability in the recent past, this paper has addressed the research question differently.

Technology Acceptance Model (TAM)

The first model used in this research for M-Government acceptance and usage is TAM. This model was proposed in 1985 to ease off the acceptability of technological services [19].

As shown in Fig. 1, the features of technological services act as a stimulus that can increase the user motivation to accept the technology. A higher explanation of the user results in higher acceptance of the technology. In TAM, user motivation can be

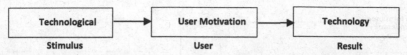

Fig. 1. Theoretical framework of the Technology Acceptance Model.

categorized into three types (i.e., perceived usefulness, ease of use, and technology attitude). In addition, these motivations can be stimulated by various variables that can belong to multiple categories like social, demographic, technological design, cultural and religious user awareness, etc. The block diagram of the more specific TAM is shown in Fig. 2.

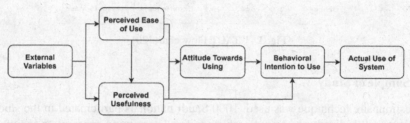

Fig. 2. Technology acceptance model for the assessment of acceptance and usability of technological services [19]

As the Fig. 2 shows, several factors have a proven impact on the adaptability of M-Government services [19]. Therefore, the effect of the above variables and other external variables of Saudi Arabia are statistically analyzed using the method of Hypothesis. In addition, Hypothesis development and testing, will be in detail in next section.

Unified Theory of Acceptance and Use of Technology (UTAUT)
It adds two more constructs as compared to TAM. These ideas help people accept social and technological change. Venkatesh et al. [20] chose this model after evaluating eight others [20]. It's a finished model. Technology acceptance researchers have thoroughly examined it [19, 20]. Based on users' behaviors'. These are the model's primary components. Performance expectation is an external variable that helps utilization. The generalized flow chart of UTAUT is visually represented in Fig. 3.

3 Methodology

This section discusses the application of technology acceptance models for assessing the acceptance of M-Government in Saudi Arabia. As various researchers have accessed technology acceptance and usability in the recent past, this paper has addressed the research question differently. Therefore, TAM and UTAUT are the most suitable technology acceptance models specifically for Saudi Arabia. These models used to study the M-Government services in Saudi Arabia. In addition, these models are customized according to the social, cultural, demographic profiles, and technological aspects of Saudi Arabia.

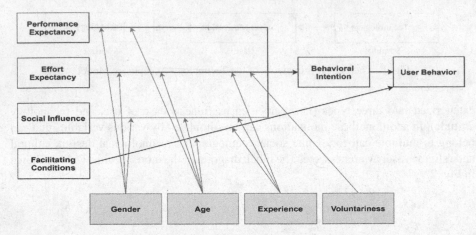

Fig. 3. UTAUT Flow chart [20].

3.1 Sample of Study

A questionnaire technique was used. 1000 Saudi nationals participated in the study as sufficient sample size based on research study for determining appropriate sample size in survey [3] (i.e., the sample represents the population). The questionnaire consists of close-ended questions. Each subject has to respond to the question on a scale of 1 (Extremely Agree) to 5 (Extremely Disagree). The questionnaire was distributed among different age groups, gender, and monthly income. Citizens living in urban, suburban, and rural areas were included. In addition, citizens with diverse educational backgrounds were considered. The questionnaire was distributed among citizens with different levels of IT experience. The questionnaire was distributed among citizens with different levels of IT experience. 60% of the subjects were male. The majority of subjects were above 30 years of age (52%). About two-thirds (73%) of issues have an intermediate level of IT literacy, and around half (55%) of subjects have an average income per month of 20,000–40,000 (SAR) as shown in Table 2.

Table 2. Demographic Profile for Participants

Gender	Age	Issues Ratio: Intermediate IT literacy	Average Income/per-month
Male (60%)	Above 30 years (52%)	About Two third (73%)	20,000–40,000(SAR) (55%)

3.2 Questionnaire

Qualtrics.com was used to send the final questionnaire to four Saudi general subject email's newsgroups for the study's full scale, which ran from early January through

February 2021. As broad subject email newsgroups, the researcher found just four major newsgroups, each with 250+ members (only Saudi citizens, Non-Saudi was excluded). Approximately 1000 people received these emails, which were used as a sample. A wide range of subjects from various interests was chosen to reflect Saudi society in these email newsgroups. After three weeks, a follow-up email was sent as a reminder, only 0.78% of those contacted completed the questionnaire (780 participants). When Neuman computed the response rate in 2006, he used this formula:

$$Response\ Rate = \frac{Total\ number\ of\ responses}{Total\ number\ of\ participants\ in\ the\ sampling\ frame}$$

The questionnaire was disseminated online to ensure that the sampled Saudi nationals have access to the Internet and hence use e-transactions; using an online sampling method would allow researchers in a large country like Saudi Arabia, as study was conducted by [21] to reach a much larger population of current or potential users of electronic transactions. A questionnaire with close-ended responses was distributed among 1000 Saudi nationals via email. Subjects have to answer each question on a scale from 1 (strongly agree) to 5 (strongly disagree). The subjects were scattered throughout Saudi Arabia, ranging from modernized urban areas to remote villages. Different age groups, gender, and IT experience were considered. The questionnaire was sent via email and hands in the villages to include highly educated personnel and illiterate citizens. The response rate of the questionnaire was 91%. The questionnaires were distributed among 60% males and 40% females. More than half the subjects were over 30 years of age. As far as the technology experience of subjects is concerned. About 73% of subjects were well aware of the internet and were intermediate-level users of internet technology. Almost 39% of subjects have never used any M-Government service in their life.

3.3 Variables and Measurement

All instruments/items were selected or used from the literature and were adapted for use in the context of m-government. Table 3 summarizes the constructs related to the proposed model UTAUT with a brief description.

Table 3. The Proposed Construct for UTAUT

UTAUT Constructs	Description
Performance Expectancy	To what extent does use believes in the successful completion of a task?
Effort Expectancy	How easily can a user achieve the goals?
Social Influence	To what extent do other citizens agree with the use in service usage?
Facilitating Conditions	To what extent, the system infrastructure and environment support users in availing the technical service?
Behavioral Intention	The user's personal chances to avail the service

4 Research Model and Hypothesis Development

As illustrated in the block diagram (Fig. 3), numerous social factors influence user's behaviour. All of these elements are taken into account while developing and testing hypotheses. For example, UTAUT's hypothesis is as follows:

H1: There is a significant positive association between users' behavioural intentions to adopt and use the M-Government system and their performance expectations.
H1-a: Gender will influence the relationship between performance expectations and users' behavioural intention to use the M-Government system.
H1-b: The link between performance expectancy and users' behavioural intention to use the M-Government system will be influenced
H2: There is a significant positive relationship between effort expectation and users' behavioural intentions to accept and use the M-Government system.
H2-a: Gender will affect the relationship between effort expectancy and behavioural intention to use the M-Government system.
H2-b: The link between effort expectancy and user behavioural intention to use the M-Government system will be influenced by age.
H2-c: Technology experience will affect the relationship between effort anticipation and user behavioural intention to utilize the M-Government system.
H2-d: The association between effort expectancy and users' behavioural intention to use the M-Government system will be influenced by voluntariness.
H3: There exists a considerably positive correlation between the influence of society and the user's behavioral intentions to accept and use the M-Government system.
H3-a: Gender will impact the correlation between the influence of society and the user's behavioural intention to use the M-Government system.
H3-b: Age will impact the correlation between the influence of society and the user's behavioural intention to use the M-Government system.
H3-c: Technology experience will impact the correlation between effort expectancy and users' behavioural intention to use the M-Government system.
H4: There exists a considerably positive correlation between conditions that facilitate the user and the user's behavioural intentions to accept and use the M-Government system.
H4-a: Age will impact the correlation between effort expectancy and user's behavioural intention to use the M-Government system.
H4-b: Technology experience will impact the correlation between effort expectancy and user's behavioural intention to use the M-Government system.

5 Result and Analysis

In this section, the statistical results of the hypothesis testing are discussed. To analyze and evaluate the statistical data, Statistical Package for Social Science (SPSS) was used.

Cronbach's Alpha formula was used to test the reliability and internal consistency to construct a model collectively. It helps to quantify the closeness of the variable set in the group of the model. Cronbach is a function of the correlation of model constructs. It is formalized by:

$$\alpha = \frac{Nc}{v + (N - 1)c}$$

N = total items.
C = the average covariance of all N elements.
V = the average variance among all N elements.

The internal consistency is excellent for value above 0.90, good between 0.90–0.80, acceptable between 0.80–0.70, questionable between 0.70–0.60, poor between 0.60–0.50 and unacceptable below 0.50. The values of Cronbach Alpha for UTAUT are given in the Table 4 below.

Table 4. Cronbach Alpha table of UTAUT Model

UTAUT Construct	# Items	Cronbach Alpha
Performance Expectancy	4	0.89
Social Influence	3	0.81
Facilitating Factors	2	0.90
Effort Expectancy	3	0.80

The table values of Cronbach show that four out of five constructs are considered good, while only one has low internal consistency. Yet, it can be categorized as acceptable. In a collectivist and tribal culture, self-enhancement ideals such as power and achievement are earned through personal connections, the energy obtained from families and supporters, and an accomplishment in leveraging social interactions to attain control or status. However, due to the disintermediation caused by online transactions, the chain of favors acquired via personal connections may be broken.

The statistical results of the hypothesis testing of UTAUT are enlisted in Table 5. Depending on the p-value, either the null hypotheses were rejected or accepted.

When items lack homogeneity, reliability criteria for reflecting constructs and objects include assessing the construct's internal consistency and delving further into the items' general correlations. Reflective things change as a result of their underlying structure changing. The surface of the group's items was assessed in 2003 by Jarvis et al. since they all represent the same concept. The detailed statistical parameters of the data collected for TAM are also shown in Table 6. Table 6 shows the mean, standard deviation, p-value, and composite reliability.

Table 5. UTAUT Hypothesis Testing Result

Hypothesis	Result (p-value)	Remarks
H1	**$\alpha = 0.49$**	**Accepted**
H1-a	Not significant	Rejected
H1 = b	0.36	Accepted
H2	**$\alpha = 0.52$**	**Accepted**
H2-a	0.33	Accepted
H2-b	Not significant	Rejected
H2-c	0.44	Accepted
H2-d	Not significant	Rejected
H3	**$\alpha = 0.33$**	**Accepted**
H3-a	0.25	Accepted
H3-b	0.41	Accepted
H3-c	Not significant	Rejected
H4	**$\alpha = 0.21$**	**Accepted**
H4-a	Not significant	Rejected
H4-b	0.50	Accepted

Table 6. Descriptive Analysis for UTAUT main Hypothesis Testing Result

Construct	Mean	*SD	P-value	#CR
H1	4.1	0..23	>0.05	0.91
H2	4.3	0.64	>0.05	0.87
H3	4.0	0.48	>0.05	0.90
H4	5.5	1.33	<0.05	0.75

* Standard deviation; #Composite Reliability

6 Discussion

The proposed hypotheses based on UTAUT were also statistically tested, and results showed that all the constructs certainly impact M-Government services acceptability. These constructs include performance expectancy, effort expectancy, social influence, and facilitating factors that affect the acceptance rate of the service provided by the Government. The impact of social influence on the Internet and the Government on the acceptance of electronic transactions was raised and addressed. In fact, these hypotheses that were addressed in the answer to the following research question. What role does social influence in the internet and in government institutions play in determining uptake of new technologies?

This research study examines the link between the social influence element and M-Government services adoption. The adoption of e-government organizations, such as M-Government apps. For instant, in Jordan is not significantly influenced by social influence, contrary to prior studies, according to the findings of this study. One benefit of social impact that Jordan may be overlooking is that when more people utilize a service and word of mouth spreads, the value of the service generally rises [22]. Likewise, it is not regarded as significant in Saudi Arabia, because there are few users have a substantial effect by social influence. It is due to the high usage of internet services (i.e. internet-societal interaction) in Saudi Arabia.

As part of determining the criteria for the study setting, several models and ideas were drawn from this research. First, a research model was built to predict and explain acceptance in the vicinity of the study. Second, an online poll was used to collect a sample of 870 Saudi residents. Third, e-transactions were studied using structural equation modelling to determine how people's willingness to use them as a communication tool with the Government interacted with their beliefs about how they work, how much faith they had in the Government, and how much social influence they had over them. Perceived compliance with values and citizens' demands and the capacity to communicate outcomes from the use of electronic transactions, faith in the internet, and conservation values are all-important positive factors in the acceptance or disapproval of e-transactions. On the other hand, social influence in government institutions and internet have a desire to attain status and control others and resources (as measured by one's "power value") have a detrimental impact on social acceptability.

7 Conclusion

M-Government adoption intentions are examined in this study, which adds to the growing body of research. Many theoretical and practical contributions are made in this study. First and foremost, this study adds theoretical depth to the TAM model by demonstrating This improved model is capable of predicting the adoption of M-Government. As a result, the model can be applied in Saudi Arabia to spot emerging patterns. For the study of M-Government technology adoption in the future. In addition, this is one of the first studies to examine the adoption of Saudi Arabia's M-Government services-based government. A country that is rapidly expanding a Middle Eastern country that is still influenced by Arab cultural and traditional values, despite the fact that it has become more modern and technologically savvy.

This study has some limitations that should be considered when interpreting the results, despite their importance. The first step is to do some research. Target homogeneity and Saudi user's perspectives restriction. The majority of respondents are university students, which may not be representative of the entire population of Saudi Arabia. From now on it will be interesting to put this system to the test on a larger scale with actual government users. Compare the findings of this study with those of other studies that publish student group brackets and other geographic categories of regions. Cross-cultural comparisons allow researchers to get the most accurate results.

References

1. Heeks, R.: Meet Marty Cooper–the inventor of the mobile phone. BBC 41(6), 26–33 (2008)
2. Rossel, P., Finger, M., Misuraca, G.: " Mobile" e-government options: between technology-driven and user-centric. Electron. J. E-gov. 4(2), 79–86 (2006)

3. Kotrlik, J., Higgins, C.: Organizational research: determining appropriate sample size in survey research appropriate sample size in survey research. Inf. Technol. Learn. Perform. J. **19**(1), 43 (2001)

4. Wang, C.: Antecedents and consequences of perceived value in mobile government continuance use: an empirical research in China. Comput. Hum. Behav. **34**, 140–147 (2014)

5. Alonazi, M., Beloff, N., White, M.: Exploring determinants of M-Government services: a study from the citizens' perspective in Saudi Arabia. In: 2019 Federated Conference on Computer Science and Information Systems (FedCSIS), pp. 627–631. IEEE (2019)

6. Almarashdeh, I., Alsmadi, M.K.: How to make them use it? citizens' acceptance of M-government. Appl. Comput. Inform. **13**(2), 194–199 (2017)

7. Althunibat, A., Abdallah, M., Almaiah, M.A., Alabwaini, N., Alrawashdeh, T.A.: An acceptance model of using mobile-government services (AMGS). CMES-comp. Model. Eng. **131**, 865–880 (2022)

8. Jasimuddin, S.M., Mishra, N., A. Saif Almuraqab, N.: Modeling the factors that influence the acceptance of digital technologies in e-government services in the UAE: a PLS-SEM Approach. Prod. Plan. Control, **28**(16), 1307–1317 (2017)

9. Abaza, M., Saif, F.: The adoption of mobile government services in developing countries. Int. J. Comput. Sci. Issues (IJCSI) **12**(1), 137 (2015)

10. Althuwaini, S., Salem, M.S.: Customer's adoption of mobile government services: the role of trust and information privacy. Int. J. Bus. Manage. Res. **9**(1), 20–27 (2021)

11. Alharmoodi, B.Y.R., Lakulu, M.M.B.: The transition from e-government to m-government: challenges and opportunities-case study of UAE. Eur. J. Multidisciplinary Stud. **5**(1), 61–67 (2020)

12. Saxena, S.: Enhancing ICT infrastructure in public services. The Bottom Line (2017)

13. Althunibat, A., Alokush, B., Tarabieh, S.M., Dawood, R.: Mobile government and digital economy relationship and challenges. Int. J. Adv. Soft Comput. Appl. **13**(1) (2021)

14. Joseph, B., Boateng, R.: Using Multiple Case Study Design to Understand the Development of Mobile Services in Ghana. SAGE Publications Ltd (2020)

15. Kim, Y., Yoon, J., Park, S., Han, J.: Architecture for implementing the mobile government services in Korea. In: Wang, S., et al. (eds.) ER 2004. LNCS, vol. 3289, pp. 601–612. Springer, Heidelberg (2004). https://doi.org/10.1007/978-3-540-30466-1_55

16. Mandari, H., Koloseni, D.: Examining the antecedents of continuance usage of mobile government services in Tanzania. Int. J. Public Adm. **45**(12), 917–929 (2022)

17. Alswaigh, N.Y., Aloud, M.E.: Factors affecting user adoption of e-payment services available in mobile wallets in Saudi Arabia. Int. J. Comput. Sci. Netw. Secur. **21**(6), 222–230 (2021)

18. Almaiah, M.A., Al-Khasawneh, A., Althunibat, A., Khawatreh, S.: Mobile government adoption model based on combining GAM and UTAUT to explain factors according to adoption of mobile government services. Int. J. Interact. Mob. Tech. (iJIM) **14**(03), 199–225 (2020)

19. Lee, Y., Kozar, K.A., Larsen, K.R.: The technology acceptance model: past, present, and future. Commun. Assoc. Inf. Syst. **12**(1), 50 (2003)

20. Venkatesh, V., Morris, M.G., Davis, G.B., Davis, F.D.: User acceptance of information technology: toward a unified view. MIS Q. **27**, 425–478 (2003)

21. Van Selm, M., Jankowski, N W.: Conducting online surveys. Qual. Quant. **40**(3), 435–456 (2006)

22. Abu-Shanab, E., Haider, S.: Significant factors influencing the adoption of m-government in Jordan. Electron. Gov. Int. J. **11**(4), 223–240 (2015)

Search by Pattern in GPS Trajectories

Maros Cavojsky$^{(\boxtimes)}$ (ID) and Martin Drozda (ID)

Faculty of Electrical Engineering and Information Technology,
Slovak University of Technology, Bratislava, Slovakia
{maros.cavojsky,martin.drozda}@stuba.sk

Abstract. In search by pattern in GPS trajectories, user draws a trajectory, the pattern query, and then receives a set of trajectories ranked by their similarity to the pattern query. We argue that when user draws a pattern query, an initial part of this query (prefix of chosen length) should have more weight than the rest of query. We assume that after receiving a set of similar trajectories, user can refine the pattern query in order to receive more relevant results. We give explanation of our approach by means of web search, where a user searches, for example, for "bratislava castle" and then adds a refinement to this query "opening hours", where removing the initial part of query does not make sense, as search for "opening hour" alone would return irrelevant results. This idea has led us to considering pattern search that is weighted toward query prefix. We experimentally evaluate this approach, in our experimentation we apply the Geolife data set (Microsoft Research Asia).

Keywords: GPS trajectory · Geolife data set · pattern search · Needleman-Wunsch algorithm · Smith-Waterman algorithm · Geohash · Hausdorff distance

1 Introduction

In search by pattern, user draws a trajectory, the pattern query, and receives a set of trajectories ranked by their similarity to this query. Preferably, the result is obtained in real time, or with negligible delay (less than 500 ms) [7]. Trajectories represent recorded positions of people or mobile platforms (cars, trucks, drones, pets etc.) equipped with GPS (Global Positioning System), or a similar system. Positions measured by GPS are estimates of real positions subject to various errors. Precision of GPS devices may be influenced by various signal propagation phenomena such as signal reflection, diffraction, scattering or multi-path propagation. In urban environments, GPS satellite occlusion may happen [13].

When comparing a query to one or several trajectories, we can view the problem as an edit distance problem, where the query and all trajectories are transformed to a letter string representation. Edit distance is the number of single letter edits that is necessary until the two representations become identical. Such a problem transformation leads to algorithms that can compute global

J. Taheri et al. (Eds.): MobiCASE 2022, LNICST 495, pp. 117–132, 2023.
https://doi.org/10.1007/978-3-031-31891-7_9

or local alignment of sequences of letters, with the Needleman-Wunsch algorithm [10] and the Smith-Waterman algorithm [16], respectively, being the prime representatives. Both these algorithms are often applied to align DNA or protein sequences. Their advantage is that they can directly align two sequences, whereby their edit distance [15], and thus their similarity can be both computed [4]. Their disadvantage is their time complexity $O(mn)$, where m is the length of the first sequence and n is the length of the other sequence. Dense urban environments give rise to an enormous number of similar trajectories, therefore applying these algorithms may be prohibitively inefficient.

When searching by pattern we can distinguish between two basic cases:

- The search returns top-k similar trajectories, where these trajectories may not have any identical sub-trajectories, or
- the search returns top-k similar trajectories, where it is required that these trajectories share a sub-trajectory, where the least length of this sub-trajectory is parameterized as γ.

In both cases it is necessary to define a similarity relation, in the latter case it is also necessary to choose the least sub-trajectory length. We will define both formally later on.

The former case leads to complex tree pruning approaches based, for example, on the Hausdorff distance [14]. An example of such an approach is the Repose approach introduced by Zheng et al. [20]. This approach relies on a trie to be computed. In order to facilitate search efficiency, this approach applies a signature based representation of trajectories based on the Z-order [5], where a $2d$-trajectory, a sequence of (x, y) positions, is transformed to a $1d$ (binary string) representation.

The latter approach, the focus of our research, assumes that there is a limited motivation for finding similar trajectories that do not share a common sub-trajectory. Similar to the previous approach, we take advantage of a $1d$ trajectory representation. Unlike in the previous case, it is not necessary to compute a trie, instead a Geohash [11] is applied to reduce the number of candidate trajectories, which are later ranked with respect to their Hausdorff distance to the query pattern.

Our results can be succinctly summarized as follow. We compare the performance of our approach to approaches applying the Needleman-Wunsch algorithm and the Smith-Waterman algorithm. For experimental evaluation we use the Microsoft Asia Geolife data set [22] that contains recorded trajectories in Beijing, China. We show that in such a dense urban environment as Beijing, the Needleman-Wunsch algorithm and the Smith-Waterman algorithm do not offer the possibility of a real time query pattern search. Therefore we propose an approach based on Geohashing that aims at real time GPS trajectory querying.

This document is organized as follow. In Sect. 2, we review the related work. In Sect. 3, we explain why we consider pattern by search that is weighted toward pattern query prefix. In Sect. 4, we introduce sequence alignment algorithms such as the Needleman-Wunsch algorithm and the Smith-Waterman algorithm. In Sect. 5, we explain our methodology for efficient search by pattern. In this section we also overview our setup for experimental evaluation. In Sect. 6, we present our experimental results, and finally, in the next section we present our conclusions.

2   Related Work

Whether the Needleman-Wunsch algorithm is suitable for pairwise trajectory comparison is in detail investigated in [3,4]. The authors point out that GPS trajectories often form clusters, also referred to as nests, that emerge due to signal reflection and multi-path signal propagation. Therefore they focus on clarifying whether this algorithm is also applicable in such detrimental conditions.

The existence of nests has attracted the attention of Yang et al. [17], where they introduce the notion of noise points. The authors suggest that one possibility for computing noise points is to compute move points and stay points, where noise points are computed as their complement.

Move Ability, proposed by Luo et al. [8], is also often applied to compute nests, or noise points. Move Ability computes the distance of end points of a GPS trajectory and compares it with the sum of distances of each successive GPS position in this GPS trajectory. Cavojsky et al. [4] apply Move Ability to filter out nests, so that GPS trajectory similarity is not affected by this phenomenon.

Zheng et al. [20] propose an approach, that they call Repose, for efficient comparison of GPS trajectories. To achieve a high degree of efficiency, they compute a trie that allows for comparing a pattern query with a set of pivot trajectories. After a necessary amount of pruning is done, a sub-set of candidate GPS trajectories is computed, the Hausdorff distance between pattern query and candidate GPS trajectories is computed, then the candidate GPS trajectory having the least distance is selected. This approach can be extended to top-k search. A deficiency of Repose is its focus on search efficiency without considering various signal propagation phenomena and their impact on measured trajectories.

Yin et al. [18] address yet another great challenge of pattern search in GPS trajectories. They propose an approach for error-bounded GPS trajectory compression, while keeping support for range queries.

The last challenge that we mention in this short literature review is GPS trajectory clustering with enhanced privacy protection of users. Zhao et. al. [19] propose to add Laplacian noise in order to achieve increased user privacy. The authors apply the Edinburgh Informatics Forum Pedestrian Database [6] that captures pedestrian traffic at the main building of the School of Informatics at the University of Edinburgh. This data set likely has no or negligible amount of noise caused by signal propagation phenomena.

3   Psychology of Search by Pattern

Let us consider the example shown in Fig. 1. It shows 5 different GPS trajectories that we want to align, and then to reason about their similarity.

While these GPS trajectories share the same direction and some of them overlap, their degree of similarity is unclear. If b becomes the pattern query, the similar trajectories should only be a and d. If a becomes the pattern query, the user would expect b as well as d as the results, since trajectories that are similar at their beginning might be *intuitively perceived* as more similar than trajectories that are similar at later trajectory stages.

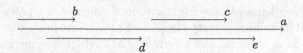

Fig. 1. GPS trajectories a,b,c,d and e to be aligned.

The rational behind giving more weight to similarity at the start rather than at the end of a trajectory is an assumption that user first starts drawing the most important portion of the search query. This then may get followed by a refinement that aims to decrease the number of possible results. This is similar to web searching, where a keyword root query is entered (for example "bratislava castle"), and then in order to obtain more relevant results, one or several keywords are added (for example "opening hours"), influenced by the results obtained in the previous search iteration. This may be considered an instance of *confirmation bias* [12], where user seeks to find a suitable result, however, then he refines the pattern query by adding to it in order to obtain a result that supports his prior believes. Notice that removing the root query ("bratislava castle") does not make sense, as searching only for ("opening hours") would return irrelevant results.

The simplest model for this kind of behavior is to discount future search keywords exponentially [1], i.e. giving less importance to refinement than to root query. This can be thus modeled as:

$$\lambda \exp^{-\lambda x},$$

where $\lambda > 0$ is a parameter of the exponential distribution, called rate. We may also require that the first γ search keywords have the same weight. This extends to a search by pattern, where user has to draw a pattern that spans at least γ different hashcodes, then a refinement can be drawn. Results are later ranked by computing their Hausdorff distance to search query.

Given two point sets A and B, the Hausdorff distance [14] between A and B is defined as:

$$H(A,B) = \max\{h(A,B), h(B,A)\}.$$

Assuming that A and B are compact sets, we can formulate h as follows:

$$h(A,B) = \max_{a \in A} \min_{b \in B} \|a - b\|, \tag{1}$$

where $\| \cdot \|$ is a norm defined on the plane. Herein we apply the Euclidean L_2 norm. The sets A and B are points in a pattern query and trajectories to which we compare.

4   Algorithms Overview

Sequence alignment is the process of aligning the letters of a pair of sequences, such that the number of matched letters is maximized. We can describe the alignment between two sequences with the following notation:

```
TACGGGCCCGCTA-C
||   | x| ||| |
TA---G-GC-CTATC
```

The vertical lines "—" (pipes), represent matching letters. Mismatches are marked by "x". Gaps are indicated by the dash "-", they are inserted between letters to keep space of missing letters in order to optimize the number of matches. In the above example, the letters A (adenine), T (thymine), G (guanine) and C (cytosine) correspond to the four known types of DNA nucleotides.

Two sequences can be aligned by computing the minimum number of single letter edits that yield these two sequences identical. The number of these edits is referred to as "edit distance" or Levenshtein distance [9]. Sellers showed that approaches formulated in terms of minimizing edit distance and maximizing similarity (e.g. Needleman-Wunch algorithm) are equivalent [15]. Wagner-Fischer algorithm [9] is often applied to compute edit distance. Wagner-Fischer algorithm, Needleman-Wunsch Algorithm and Smith-Waterman Algorithm are examples of dynamic programming algorithms.

4.1   Needleman-Wunsch Algorithm (NWA)

NWA is a global alignment algorithm, meaning the result always aligns the entire input sequences. The algorithm is using scoring matrix for its sequence alignment, in simpler case defined as:

$$S_{a,b} = \begin{cases} 1, & \text{if } a = b, \\ -1, & \text{if } a \neq b. \end{cases}$$

In addition to a scoring matrix, algorithm also applies penalties for gaps. The most common gap penalty is the linear gap penalty, defined as follows:

$$W_L(d) = Gd,$$

where the gap penalty W_L is proportional to the length d of the gap by a parameter G. There are cases where more complicated approach is necessary, for example, an "affine gap penalty" penalizes opening a gap by one parameter, and extending the gap by another parameter. Such a gap penalty can by defined as follows:

$$W_L(d) = G + (d - 1)E,$$

where we include a gap open penalty G and a gap extension parameter E proportional to the length d of the gap.

The algorithm computes the score matrix F using a recurrence relation, such that the values of a given cell of the matrix F are defined in terms of the neighboring cells. NWA applies for computing an alignment of two sequences x and y the following recurrence relation:

$$F_{i,j} = max \begin{cases} F_{i-1,j} + G, & \text{skip a position in x,} \\ F_{i,j-1} + G, & \text{skip a position in y,} \\ F_{i-1,j-1} + S_{x[i],y[j]}, & \text{match / mismatch.} \end{cases}$$

4.2 Smith-Waterman Algorithm (SWA)

In many cases we are only interested in aligning a portion of the sequence to find a local alignment. Furthermore, we do not necessarily want to force the first and last letter to be aligned. SWA is an alignment algorithm that has these properties. We can define a set of boundary conditions for the scoring matrix $F_{i,j}$, namely that the score is 0 at the boundaries:

$$F_{i,0} = 0, \text{ for } 1 \leq i \leq \text{length}(x),$$

$$F_{0,j} = 0, \text{ for } 1 \leq i \leq \text{length}(y).$$

For the rest of score matrix we apply the following recurrence relation:

$$F_{i,j} = max \begin{cases} F_{i-1,j} + G, & \text{skip a position in x,} \\ F_{i,j-1} + G, & \text{skip a position in y,} \\ F_{i-1,j-1} + S_{x[i],y[j]}, & \text{match / mismatch,} \\ 0, & \text{zero-out negative scores.} \end{cases}$$

The key difference between NWA (global alignment) and SWA (local alignment) is that when computing the global alignment we start backtracking from the lower right cell of the matrix. To compute the local alignment we instead start at multiple cells which all share the maximum score in matrix and we keep backtracking until we reach zero score.

5 Methodology

In our experimental evaluation we investigate the suitability of three approaches for comparing trajectories:

- our approach based on geohashing,
- Needleman-Wunsch Algorithm (NWA), and
- Smith-Waterman Algorithm (SWA).

All three approaches apply the same set of search patterns and all experiments are based on the Geolife data set.

The Geolife data set (Microsoft Research Asia) was collected by 182 users in a period of over three years (from April 2007 to August 2012). This data set contains 17,621 trajectories with a total distance of about 1.2 million kilometers and a total duration of 48,000+ h. These trajectories were recorded by different GPS loggers and GPS capable phones, and have a variety of sampling rates. This data set recorded a broad range of users' outdoor movements, including not only life routines such as "go home" and "go to work" but also entertainment and sports activities, such as shopping, sightseeing, dining, hiking, and cycling, for details see [21–23].

Several relevant statistics for the Geolife data set are shown in Figs. 2(a) and 2(b); see [2] for additional useful statistics.

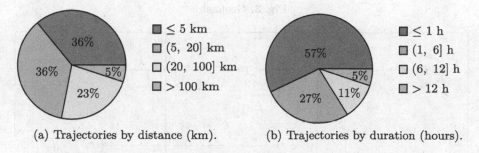

(a) Trajectories by distance (km).   (b) Trajectories by duration (hours).

Fig. 2. Distribution of trajectories in Geolife data set.

5.1   Pre-processing Geolife Data Set

One of the options how to pre-process trajectories is using Geohash [11], a hierarchical spatial data structure, that encodes a GPS location into a string of digits and letters. Encoding is done with subdividing space into buckets of grid shape, called Z-order curve. Latitude and longitude is encoded by base32, that uses digits 0–9 and almost all lower case letters except "a", "i", "l" and "o". If geohashes share the same prefix, the points within these geohashes are spatially close. The opposite may not hold, since points can be close to each other but may not share a prefix. An example of Geohash grid is shown in Fig. 3.

By using Geohash we are able to assign to every trajectory a set of geohashes that contain the points that belong to these trajectories. Before searching by pattern, we calculate the set of geohashes that represent the search pattern. We then select trajectories which have a non-empty intersection with this set of geohashes.

Let us consider the example depicted in Fig. 4, where solid circles are recorded locations and dotted circles are calculated locations by linear interpolation to fill geohashes between locations with gaps.

Pre-processing then consists of these steps:

Fig. 3. Geohash.

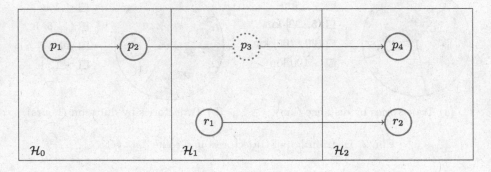

Fig. 4. Trajectories p and r encoded with geohashes.

- All positions are transformed to geohashes. We use geohashes with the length of 7, however, any granularity can be achieved by using shorter or longer geohashes.
- Repeatedly occurring geohashes are deleted, i.e. the next geohash in the series is always different from the previous one.
- If necessary, missing geohashes between positions are calculated by linear interpolation. If we do not include geohashes for gaps, our approach is unable to compare trajectories; see [4] for more information on gaps and on how they emerge.
- We do not consider trajectories that are both short and require interpolation. Interpolated geohash trajectories with length less then 6 geohashes are not considered in our evaluation.

As already described above, when applying NWA, SWA or our approach for similarity, we run these approaches on tracks, not on original GPS trajectories that we use to compute tracks.

Definition 1. *Let $\mathcal{H}_i$ be the i-th geohash, id the track identifier, and ℓ the track length. Track t is then defined as a sequence of pairs $t = [(\mathcal{H}_0, id), (\mathcal{H}_1, id), \cdots , (\mathcal{H}_{\ell-1}, id)]$, where $\mathcal{H}_i$ and $\mathcal{H}_{i+1}$ are neighboring geohashes.*

5.2 Our Approach Based on Geohashing

To search similar tracks with geohashing we first assign tracks to a hash map. We create hash keys by concatenating every γ geohash codes in track. Each hash key in hash map points to a set of tracks ids, that share that tuple of geohash sequence. The considered approach is shown in Fig. 5, where the tracks t_1, t_2 and t_3 are assigned to four hash map keys.

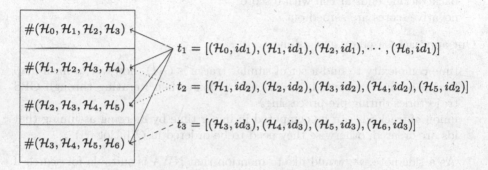

Fig. 5. Hashing tracks based on geohash codes for $\gamma = 4$.

In order to find tracks that are similar to search query $q = [(\mathcal{H}_0, id_0), \cdots, (\mathcal{H}_4, id_0)]$, the following steps get applied:

- Split q to tuples by applying a moving window of the size γ that at each iteration moves to the right by one geohash.

  For example for a track of length 5, and $\gamma = 4$ we compute two tuples (hash map keys):
  $[(\mathcal{H}_0, \mathcal{H}_1, \mathcal{H}_2, \mathcal{H}_3), (\mathcal{H}_1, \mathcal{H}_2, \mathcal{H}_3, \mathcal{H}_4)]$.

- Retrieve tracks from keys of hash map based on tuples from pattern, and compute the union of retrieved tracks.

  For example:
  $r = \{id_1, id_2\} \cup \{id_1, id_2, id_3\} = \{id_1, id_2, id_3\}$
- Rank r by similarity to q, applying the Hausdorff distance between q and each result in r.

5.3 Comparison of Methods

The approaches considered herein can be succinctly characterized as follow:

Needleman-Wunsch Algorithm (NWA)

- time complexity to compute optimal global alignment(s) is $O(mn)$

- backtracking starts only in lower right cell
- backtracking ends in upper left cell
- allows for negative scores

Smith-Waterman Algorithm (SWA)

- time complexity to compute all local alignments is $O(mn)$
- backtracking starts at cells with maximum score
- backtracking ends at cell with 0 score
- negative scores are zeroed out

Our approach based on geohashing

- time complexity to find a set of similar tracks is $O(1)$, however, a hash map has to be first computed, which can be done when iterating through GPS trajectories during pre-processing
- union of track ids can be computed in linear time by merging assuming that ids are ordered, otherwise they need to be ordered in $O(|r|\log|r|)$

As a side note, we would like to mention that NWA is suitable for search if we expect trajectories to be nearly identical, on the other hand, SWA is suitable when we expect trajectories to be similar at least in some segment. How our approach based on geohashing compares to NWA and SWA is one of the goals of our experimental evaluation.

6   Experimental Results

For searching in the Geolife data set, we apply two pattern queries shown in Figs. 6 and 7. They both run through the city center of Beijing. The former one is a short query with the length of 11 km and the latter is a trip from the airport to the city center, its length is 39 km. The long pattern query contains the short pattern query in order to simplify reasoning about obtained results.

6.1   Searching with Needleman-Wunsch Algorithm

For searching using NWA we apply the parameters specified in Table 1, these parameters are also applied in a previous study on trajectory similarity by Cavojsky and Drozda [3]. The similarity between two tracks is therein defined as follows:

Definition 2. *Let C_a and C_b be two tracks. Tracks C_a and C_b are similar if these tracks have at least α matches and no more than β subsequent mismatches.*

We run NWA incrementally by increasing the length of pattern query by 5 Geohash codes (approx. 600 m) up to the full length of search pattern for both the short and long pattern. In other words, we start with a prefix of the pattern query having the length of 5 Geohashes, subsequently we increase its length by 5

Fig. 6. Short pattern with the length of 11 km (starts on the right).

Fig. 7. Long pattern with the length of 39 km (starts on the right).

Table 1. Needleman-Wunsch Algorithm parameters

match score	1
mismatch penalty	−1
gap penalty	0
α	$\gamma = 4$
β	3

Geohashes, until full pattern query length is achieved. This is necessary in order to evaluate the scalability of NWA for different query lengths.

Figures 8(a) and 8(c) show the results for these two search patterns. We can see that as the length of either search pattern increases, so does increase the time to compute similar tracks in the Geolife data set. We consider these results to be unsuitable for real time search, since they are beyond what can be considered as "negligible delay", where Liu and Heer [7] show that a delay of 500 ms already results in decreased user activity (in use cases with high interaction).

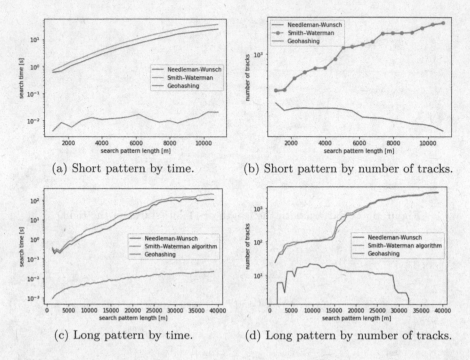

(a) Short pattern by time.   (b) Short pattern by number of tracks.

(c) Long pattern by time.   (d) Long pattern by number of tracks.

Fig. 8. Comparison of different methods based on search time and found tracks.

Figures 8(b) and 8(d) show the number of similar track. As expected, the number of these tracks decreases with the pattern query length.

6.2   Searching with Smith-Waterman Algorithm

For searching using Smith-Waterman Algorithm we used parameters specified in Table 2. We set mismatch and gap penalties to –1, because our goal is to search for maximal alignment. If two tracks have at least γ subsequent matches we call them similar.

Table 2. Smith-Waterman Algorithm parameters.

match score	1
mismatch penalty	–1
gap penalty	–1
subsequent matches	$\gamma = 4$

The experiment is done the same way as in the case with NWA, i.e. the length of pattern query is incrementally increased in order to test with still longer pattern queries.

Figures 8(a) and 8(b) show the results for the two pattern queries. We can again see that the time to compute an alignment cannot be considered negligible. Figures 8(b) and 8(b) show the number of found tracks.

6.3   Searching with Geohashing

We applied the same search procedure as for NWA and SWA for both the short and long pattern. Figures 8(a) and 8(c) show the time needed to compute r, i.e. the set similar tracks, that still need to get ranked by its similarity to pattern query. We can see that compared to both NWA and SWA, the search time is considerably decreased. We can consider these results as having negligible delay, thus suitable for real time search. Figures 8(a) and 8(c) show the number of similar tracks.

The results for the search by pattern for NWA, SWA and our approach based on geohashing shown in Fig. 8 indicate that the latter approach is a dominant approach in terms of computational time. The results also indicate that SWA and our approach based on geohashing are similar in terms of what they find, i.e. local alignments. Notice that our experimental results are based on a data set with a certain number of measured GPS positions, therefore we are unable to provide statistical significance for our results. We only have two samples for each search pattern length. We are currently focusing on creating a synthetic data set that could resolve this problem [2], however, as a proof of concept the results give a clear direction for future research.

6.4   Search by Pattern Starting with γ Identical Geohashes

We have argued that a general pattern search that compares any search pattern to a set of tracks with equal weight may not be necessary. Therefore we suggested that any found tracks need to start with γ geohashes that are identical to initial γ geohashes of the search pattern.

In order to compare the general approach and our approach based on γ identical geohashes, we compute the Hausdorff distance for the results obtained by these two approaches. These results are shown in Figs. 9(a) and 9(b) for the general approach and in Figs. 9(c) and 9(d) for our approach. We can see that requiring that initial γ geohashes are identical has a great potential to further decrease computational burden. Notice that the computed Hausdorff distances may seem large (500m to tens of kilometers), however, the Hausdorff distance is computed as the maximum minimum distance between two sets of points, rather than an average or minimum distance; see Eq. 1 for details. In the case of the long pattern shown in Fig. 9(d), the number of reasonable results has shrunk to just three possibilities.

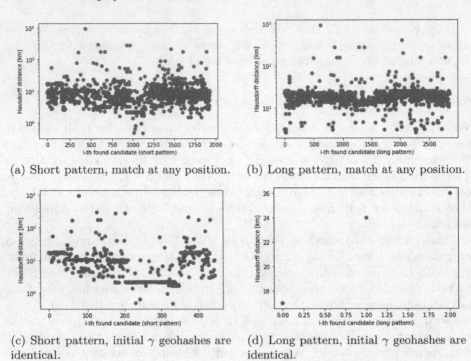

(a) Short pattern, match at any position.   (b) Long pattern, match at any position.

(c) Short pattern, initial γ geohashes are identical.   (d) Long pattern, initial γ geohashes are identical.

Fig. 9. Hausdorff distance for considered tracks using geohashes.

7 Conclusion

We argue that if search by pattern in GPS trajectories is similar to web search, user first draws the most important part of the query, and then influenced by the result, a query refinement may get added. This idea has led us to designing for a specific type of pattern search, where it is not necessary to compare search query with a potentially very large set of recorded trajectories. We pointed out that such a general approach might lead to computing a trie associated with complex pruning rules.

Instead we investigate the option where initial part of pattern query has more weight than any possible tail. We prepare several experiments based on the Geolife data set that contains GPS trajectories recorded in Beijing, China. We show that giving the initial part of a pattern query more weight has a potential to improve query time and thus how user might interact.

There are several challenges and deficiencies of our approach that we would like to stress. Even though the Geolife data set includes GPS trajectories recorded in a major urban area, it does not by far capture the level traffic that happens in extremely dense urban areas. We are currently working on an approach that could generate large scale synthetic data sets of GPS trajectories [2]. A massive data set could then give a final result on efficiency of various search approaches.

Drawing a pattern query is a major challenge since it requires a well-designed user interface that allows for simple pattern drawing, its refinements and adjustments. As simple as it sounds, it is a quite challenging task connected with a lot of effort aimed at user experience testing.

References

1. Baker, R.D.: Mathematical models of confirmation bias (2022). https://doi.org/10.48550/ARXIV.2202.03072, https://arxiv.org/abs/2202.03072
2. Cavojsky, M., Drozda, M.: Gap analysis of cohave, geolife and T-Drive datasets. In: 14th International Scientific Conference on Distance Learning in Applied Informatics (DIVAI), pp. 356–364. Wolters Kluwer (2022). https://www.divai.sk/assets/divai2022.pdf
3. Čavojský, M., Drozda, M.: Comparison of user trajectories with the needleman-wunsch algorithm. In: Yin, Y., Li, Y., Gao, H., Zhang, J. (eds.) MobiCASE 2019. LNICST, vol. 290, pp. 141–154. Springer, Cham (2019). https://doi.org/10.1007/978-3-030-28468-8_11
4. Čavojský, M., Drozda, M., Balogh, Z.: Analysis and experimental evaluation of the Needleman-Wunsch algorithm for trajectory comparison. Expert Syst. Appl. **165**, 114068 (2021). https://doi.org/10.1016/j.eswa.2020.114068
5. Dai, H.K., Su, H.C.: On the locality properties of space-filling curves. In: Ibaraki, T., Katoh, N., Ono, H. (eds.) ISAAC 2003. LNCS, vol. 2906, pp. 385–394. Springer, Heidelberg (2003). https://doi.org/10.1007/978-3-540-24587-2_40
6. Fisher, B.: Edinburgh informatics forum pedestrian database (2010). https://homepages.inf.ed.ac.uk/rbf/FORUMTRACKING/. Accessed 11 July 2022
7. Liu, Z., Heer, J.: The effects of interactive latency on exploratory visual analysis. IEEE Trans. Visual Comput. Graphics **20**(12), 2122–2131 (2014)
8. Luo, T., Zheng, X., Xu, G., Fu, K., Ren, W.: An improved DBSCAN algorithm to detect stops in individual trajectories. ISPRS Int. J. Geo Inf. **6**(3), 63 (2017)
9. Navarro, G.: A guided tour to approximate string matching. ACM Comput. Surv. **33**(1), 31–88 (2001). https://doi.org/10.1145/375360.375365
10. Needleman, S.B., Wunsch, C.D.: A general method applicable to the search for similarities in the amino acid sequence of two proteins. J. Mol. Biol. **48**(3), 443–453 (1970)
11. Niemeyer, G.: Geohash. http://geohash.org/. Accessed 13 July 2022
12. Plous, S.: The Psychology of Judgment and Decision Making. Mcgraw-Hill Book Company, New York (1993)
13. Rappaport, T.S., et al.: Wireless Communications: Principles and Practice, vol. 2. Prentice Hall PTR, New Jersey (1996)
14. Rucklidge, W. (ed.): The hausdorff distance, vol. 1173, pp. 27–42. Springer, Heidelberg (1996). https://doi.org/10.1007/BFb0015093
15. Sellers, P.H.: On the theory and computation of evolutionary distances. SIAM J. Appl. Math. **26**(4), 787–793 (1974)
16. Smith, T., Waterman, M.: Identification of common molecular subsequences. J. Mol. Biol. **147**(1), 195–197 (1981). https://doi.org/10.1016/0022-2836(81)90087-5
17. Yang, Y., Cai, J., Yang, H., Zhang, J., Zhao, X.: Tad: a trajectory clustering algorithm based on spatial-temporal density analysis. Expert Syst. Appl. **139**, 112846 (2020). https://doi.org/10.1016/j.eswa.2019.112846

18. Yin, H., Gao, H., Wang, B., Li, S., Li, J.: Efficient trajectory compression and range query processing. World Wide Web **25**(3), 1259–1285 (2022)
19. Zhao, X., Pi, D., Chen, J.: Novel trajectory privacy-preserving method based on clustering using differential privacy. Expert Syst. Appl. **149**, 113241 (2020). https://doi.org/10.1016/j.eswa.2020.113241
20. Zheng, B., Weng, L., Zhao, X., Zeng, K., Zhou, X., Jensen, C.S.: Repose: distributed top-k trajectory similarity search with local reference point tries. In: 2021 IEEE 37th International Conference on Data Engineering (ICDE), pp. 708–719. IEEE (2021)
21. Zheng, Y., Li, Q., Chen, Y., Xie, X., Ma, W.Y.: Understanding mobility based on GPS data. In: Proceedings of the 10th International Conference on Ubiquitous Computing, pp. 312–321. ACM (2008)
22. Zheng, Y., Xie, X., Ma, W.Y.: Geolife: a collaborative social networking service among user, location and trajectory. IEEE Data Eng. Bull. **33**(2), 32–39 (2010)
23. Zheng, Y., Zhang, L., Xie, X., Ma, W.Y.: Mining interesting locations and travel sequences from gps trajectories. In: Proceedings of the 18th International Conference on World Wide Web, pp. 791–800. ACM (2009)

Author Index

J. Taheri et al. (Eds.): MobiCASE 2022, LNICST 495, p. 133, 2023.
https://doi.org/10.1007/978-3-031-31891-7

Printed in the United States
by Baker & Taylor Publisher Services